Geologic History of the Feather River Country, California

Geologic History of the Feather River Country, California

CORDELL DURRELL

University of California Press

Berkeley Los Angeles London

This book is a print-on-demand volume. It is manufactured using toner in place of ink. Type and images may be less sharp than the same material seen in traditionally printed University of California Press editions.

The paper used in this publication meets the minimum requirements of ANSI/NISO Z39.48-1992 (R 1997) (Permanence of Paper).

University of California Press
Berkeley and Los Angeles, California

University of California Press, Ltd.
London, England

Library of Congress Cataloging-in-Publication Data

Durrell, Cordell.
Geologic history of the Feather River Country, California.

Bibliography: p.
Includes index.
1. Geology—California—Feather River Region. 2. Geology—Sierra Nevada Mountains (Calif. and Nev.)
I. Title.
QE90.F4D87 1986 557.94'3 85-24522
ISBN 0-520-05261-7

This book is dedicated
to the memory of
Helen Margaret Spinning Durrell,
1914–1983,
whose love and devotion made
it possible.

Contents

Preface

I have come to have a great affection for the country whose geologic history is related in this book, especially for the chain of valleys between Calpine and Quincy, and most of all for the Mohawk Valley, which was my place of temporary residence during most summers from 1938 to 1977. It is a region of great beauty as well as of fascinating geology, and I hope through this account to share with the public the enjoyment that I have had there over the years. I have used a minimum of technical jargon so that the material may be understandable and interesting to all, especially those who may not have a background in geology.

When I first began to plan this book it was to be a description of the geology of a small area in and around the Plumas-Eureka State Park, which is most interesting for its geology, its scenery, and its human history. The park contains the now exhausted Jamison and Plumas-Eureka lode mines, once highly productive underground workings for the recovery of gold from veins of quartz; relicts of extensive placer diggings along Jamison and Squirrel creeks; and the ghost town Johnsville, which served the mines when they were active. The evidences of geologic history, both very old and very recent, are remarkably well displayed. However, it soon became clear that to describe well that small area required reference to a larger area, and thus the subject expanded into the geology of the northern Sierra Nevada—that part of the range north of Interstate 80. Yet for the most part this account focuses on Plumas, Sierra, and Butte counties, with lesser references to Lassen County and the counties south of the North Fork of the Yuba River, essentially the Feather River country.

My interest in the Feather River country began early in my life, in 1921 to be exact, when I accompanied my parents on what started to be a camping trip to the Lassen Peak area. We reached Manzanita

Lake from Redding with the help of considerable pushing by my mother and myself while my father slipped the clutch as an aid to navigation over the roughest and rockiest parts of what passed for a road. At Manzanita Lake we found that our car could go no farther toward the area devastated by the eruptions of Lassen Peak from 1914 to 1917 and we were forced to return to Viola, where we turned south to Mineral and then east to Chester. There my parents decided to go to Lake Tahoe by way of the route now designated as State Highway 89. At that time there were only county roads, unpaved, and only locally improved over what they had been in the days of horse-drawn coaches and freight wagons. We had left pavement at Redding and did not see it again until we neared Sacramento.

From Chester we passed Lake Almanor, then filled with standing dead trees for the forest had not been cleared before the reservoir was filled. We went through Indian Valley and down Indian Creek by a truly frightening one-way road, the remains of which can still be seen well above the present highway between Arlington Bridge and Indian Falls, and then on to Quincy. From Quincy we continued on the county road, now only roughly paralleled by State Highway 89, although the remains of that road can still be identified in many places. We camped overnight in the Mohawk Valley near the sawmill that is now Graeagle and then went on to Truckee and Lake Tahoe. On the grade between Clio and Calpine, then no more than a pair of wheel tracks, my mother and I again pushed with vigor and ate a considerable amount of red dirt while doing so. Such was my introduction to the northern Sierra Nevada, the Feather River country, and Plumas County.

I did not return until the summer of 1938, when a new topographic map, the Blairsden quadrangle, issued by the U.S. Geological Survey, became available. By then I was an instructor in geology in the University of California, Los Angeles, and I was seeking a place to continue research in the Sierra Nevada that I had begun in 1932 in Tulare County.

The area covered by the Blairsden map looked very promising. It is ideally situated for a study of the east front of the Sierra Nevada for its southwest corner is on the crest of the range, and it extends to the northeast across the Mohawk Valley and Grizzly Ridge. I returned the next summer to start work in earnest, and I have not yet completed it to my satisfaction.

As is commonly the way in geological studies, it soon became clear that the solution to problems arising in the Blairsden quadrangle required the study of adjacent areas also, and so, in later years with the aid of graduate students, my interests expanded to the north into Indian Valley, west to Quincy, east as far as Pyramid Lake, and southwest beyond La Porte. A multitude of problems have been solved and many remain to be solved.

Some of the geological studies carried out in this region since I began work in 1939 have been published, but much of that done by graduate students has not, although the theses submitted are available in the University of California libraries at Los Angeles and Davis. I have drawn freely upon those studies in preparing this manuscript, as I have also upon earlier works, especially those of J. S. Diller, W. Lindgren, and H. W. Turner, all of the U.S. Geological Survey, who were active in the period 1880–1910.

Although this report contains a number of maps, not all places named or locations described by Section, Township, and Range are noted on them. It will be well for the reader to have a copy of the excellent Forest Visitor's Map of the Plumas National Forest, available at local district ranger stations, the Plumas National Forest headquarters in Quincy, and the Forest Service's regional headquarters in San Francisco. This map shows the U.S. Public Survey lands, all passable roads, as well as some that may not be passable, and all place names that are in common use.

Very useful also are the topographic quadrangle maps of the United States Geological Survey (USGS) at scales of 1 inch to the mile (15-minute series), 1 inch to 2,000 feet (7.5-minute series), and 1 inch to 4 miles (1-degree-by-2-degree series). Businesses that sell them can be located in the telephone directory under Maps.

Sheets of the Geologic Map of California are available at the offices of the California Division of Mines and Geology in Sacramento, Pleasant Hill, and Los Angeles. The Chico, Sacramento, and Westwood sheets pertain to the northern Sierra Nevada. These excellent and useful colored maps are printed over the 1-by-2-degree USGS topographic sheets of the same names. Unfortunately, the color patterns tend to conceal the roads and place names.

I have not documented this work in the manner customary in scholarly publications for the reason that it seems an unnecessary complication in an account intended to be read by a public whose interests are not likely to cause them to seek out the original literature. I have, however, noted the names of original workers at appropriate places in the text, and I have listed in the bibliography all the sources that I have drawn upon.

I think that no work in geology based, as this is, on field studies is ever finished or can be perfect. I feel that I have not finished what I began—I wish I had another twenty years to continue to examine this fascinating region. However, I accept responsibility for errors of omission and commission that will undoubtedly come to light in the future.

I have had assistance from many people. Foremost among them was my beloved and uncomplaining wife, Helen, who camped out with me in the early years and slept on the ground before we could afford the luxury of air mattresses—those were the depression years. Later she spent her summer "vacations" cooking and keeping house just as she did during the rest of the year but without the appurtenances and conveniences of home, feeding not only ourselves but also geologists passing through and especially hungry students, most of whom were camping out, who came not only for consultation about their problems but also for food, refreshment, and, after we could afford a house, a hot shower.

During my years of semiresidence in Plumas County I made many friends among the local residents, many of them now passed on. Among them were the superintendents of the Plumas-Eureka State Park and the district rangers of the U.S. Forest Service at the Mohawk Ranger Station. Although too numerous to mention by name, I remember them all fondly and I thank those who remain for their friendship and assistance. I am especially grateful to Mrs. Ellen Bailey Guttadauro and Mrs. Mary Graziose for their skillful assistance with the drawings and photographs, respectively.

My work has been financed partly by myself and partly by funds provided by the Committees on Research of the Academic Senate at the University of California, Los Angeles and Davis.

PART I

An Introduction to the Sierra Nevada

1

Some Notions About Geology and Especially About Rocks

This chapter is a brief introduction to the science of geology that will aid the reader in understanding what follows. It contains something about the principles upon which geologists are agreed, something about the processes that affect the earth, and something about the materials of which the visible part of the earth is composed. I have illustrated the chapter as far as possible with examples from the Feather River country, so that one may visit the places illustrated to see what things really look like, for photographs and drawings never adequately portray the original.

The earth is layered as shown in fig. 1. It consists of a very dense central core, most probably of metallic iron and nickel, surrounded by a less dense mantle of stony material. The uppermost part of the mantle between depths of 42–155 miles (70–250 km) below the surface is a weak plastic layer called the asthenosphere, which is in turn surrounded by the rigid material of the lithosphere. The lithosphere includes the crust, which is composed of rocks like those visible at the surface. But the crust is of two different kinds: that of the continents is rock of low density that has an average composition about like that of granite and is about 30 miles (50 km) thick; that of the ocean basins is heavier, composed mostly of the dark-colored volcanic rock called basalt and is about 6 miles (10 km) thick. Basalt also underlies the lighter rock of the continents. Figure 2 shows the crust and the major relief of the earth's surface.

The crust is divided into about a dozen large "plates" and a number of small ones, most of them consisting of both continental and oceanic crust. The plates move slowly over the earth, permitted to move by plastic flow in the underlying asthenosphere. The motion of plates is associated with the creation of new oceanic crust by volcanic

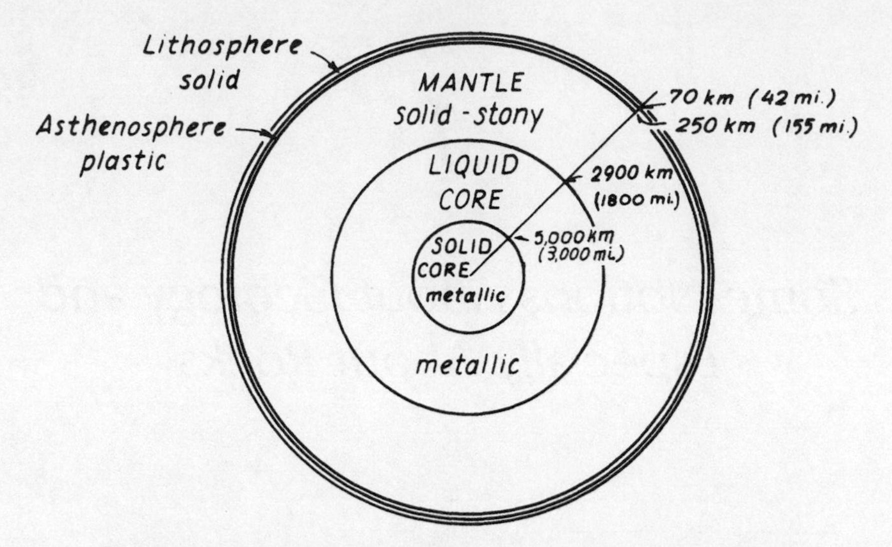

Fig. 1. The interior of the earth. A solid central core is surrounded by a liquid zone, both probably of metallic iron and nickel. The core is surrounded by the mantle of stony material. That in turn is overlain by the thin asthenosphere, which is plastic, and by the solid lithosphere or crust.

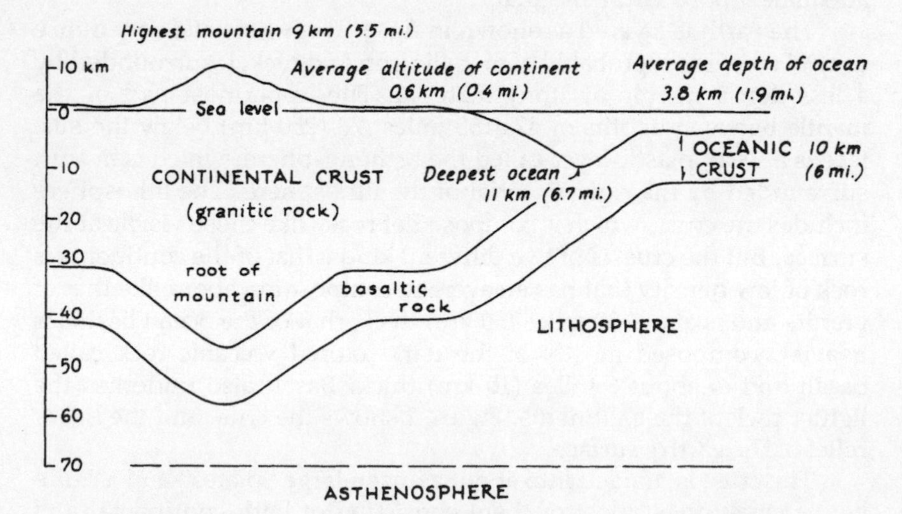

Fig. 2. The major relief of the earth's surface, and the relationship of the continental and oceanic crusts to each other and to the lithosphere and the asthenosphere.

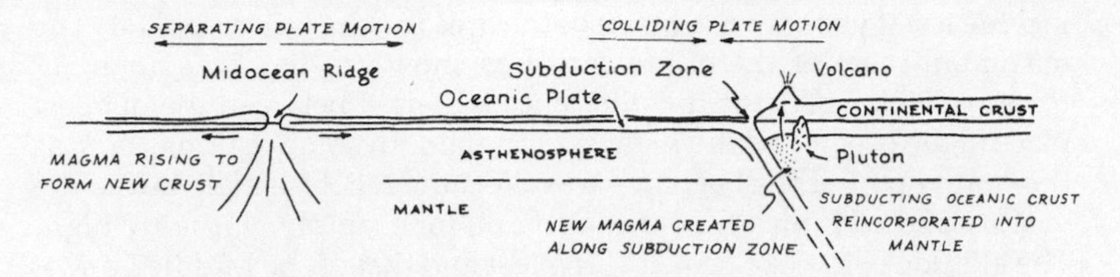

Fig. 3. Illustration of the plate tectonic concept. Plates originate at midocean ridges by the intrusion of basaltic magma (molten rock) derived from the mantle below. The plates move away from the midocean ridge, perhaps under the influence of movement in the mantle. At the subduction zone the plates are carried down, or "subducted," and restored to the mantle. Melting along the subduction zone gives rise to volcanoes and intrusions of coarse-grained igneous rock called plutons in the upper plate. The upper plate may be either continental crust, as shown, or oceanic crust.

activity along the midocean ridges, where plates separate and move away from each other. At the opposite edges of plates the oceanic crust turns downward and is reincorporated into the mantle (fig. 3). This is a very brief and simplified description of "plate tectonics," a new concept but one now accepted by nearly all geologists. Because the continents move with the plates the older term "continental drift" is also applied to the concept. More is said about plate tectonics further on.

The crust of the earth is surrounded by a shell of gas, the atmosphere. Mixed with the crust and the atmosphere is the hydrosphere, which includes the oceans, lakes, streams, underground waters, the ice of glaciers, and the water vapor in the atmosphere.

The crust, atmosphere, and hydrosphere interact with each other to produce, among other things, soil, a mixture of decayed rock and decayed and living organic matter. The thin and fragile soil layer, together with water of the hydrosphere and the oxygen and carbon dioxide of the atmosphere, supports virtually all life on earth above sea level. It supports the trees of the forest, the shrubs of the brushland, the smaller plants of the meadow and other open places, and indirectly all terrestrial animal life. The soil, the forest, and the brush tend to conceal from the eye the nature and arrangement of the rocks of the crust. But in mountainous lands the soil cover is thin in most places, and nonexistent in others, so that it is not a great obstacle to an understanding of the history recorded in the rocks below.

The surface of the solid earth is irregular; it is said to have relief, expressed as differences in altitude. Any statement about relief must include a statement of the limits of the area to which it is applied. The maximum relief of the entire earth, as shown in fig. 2, is about 12 miles, or 20 km. That is the difference in height between the deepest place in the ocean and the highest mountain. In some regions, such as the central part of the Sacramento Valley, the relief is small, but in the vicinity of the Plumas-Eureka State Park, for example, it is fairly high. The altitude of Eureka Peak is 7,286 feet and that of the Middle Fork of the Feather River near Mohawk is about 4,300 feet, so the local relief is about 3,000 feet. The relief of the earth is the result of two sets of forces: (1) those that are internal, or that work from within the solid earth, and (2) those that are external, that result from the interactions of the atmosphere, the hydrosphere, and the lithosphere.

Major features of relief such as mountain ranges, volcanoes like Mt. Shasta, and valleys such as the Sacramento and Mohawk, result from internal processes. Most smaller features of relief result from external processes.

WEATHERING

Weathering is the result of chemical reactions of water and the atmospheric gases oxygen and carbon dioxide with the rocks of the lithosphere. It is a process of decay or decomposition. Hard rocks become cracked and fragmented. Mineral constituents of the rock, originally hard and vitreous in appearance, lose their luster, are softened, and eventually turn into clay, which is often stained buff, brown, or rust colored by iron oxides. The iron oxide comes from iron-bearing minerals that were originally green or black. By decay the rock slowly becomes a soft, incoherent, or loose aggregate of clay, iron oxide, rock fragments, partly decayed rock, and grains of minerals, such as quartz, that are resistant to decay. When dry the mass may appear powdery; when wet it may be mostly mud. Such weathered rock is a part of the soil, but in most places the decay process extends well below material commonly called soil. Transitions from rock to soil (the weathering profile) can be seen in almost any road cut, as in fig. 4, which shows a dark-colored root-filled soil above the nearly white volcanic ash of the Delleker Formation, from which it was derived. In contrast, fig. 5 shows the granular product of decay of coarse-grained granitic rock. The loose material at the base of the slope is called grus by geologists and decomposed granite by everyone else. That not all rocks decay at the same rate is indicated by fig. 6, which shows the wall-like outcrop of a dike of hornblende andesite that stands far above the hornblende andesite mudflow breccia that it intruded because it is much more resistant to decay.

Fig. 4. Soil, dark colored, stony, and filled with roots, over the nearly white rock of the Delleker Formation, from which it is derived. Along the Mohawk logging road, Sec. 2, T. 21 N., R. 12 E.

Fig. 5. Deeply weathered granitic rock. On State Highway 89 between Graeagle and Calpine.

Fig. 6. Differential weathering. A dike of andesite stands like a wall above the surrounding, more easily weathered, mudflow breccia of the Bonta Formation. The dike has columnar jointing. The columns are horizontal, elongate perpendicular to the sides of the dike, and resemble stacked cordwood. By the road from the west side of Lake Davis to Smith Peak. Near the SW Cor., Sec. 4, T. 23 N., R. 13 E.

EROSION

Weathered material, softened, and with reduced coherence, is subject to erosion—that is, movement and wear by moving water, wind, or ice. Once in transit the solid particles are called sediment, and are subject to change as a consequence of the transportation process. Particles the size of sand grains and larger become rounded by wear unless the distance and time of transport are very short. But some fragments, especially those transported by ice, as explained more fully in Chapter 8, take on angular shapes that are a characteristic result of that process.

Not only are particles of rocks and minerals worn in transport, they also wear away the rock of the surfaces across which they are transported. Windblown sand wears away hard rock in the same way that the industrial process of sandblasting removes old paint, etches wood, or removes rust from steel. Sand and larger particles—pebbles, cobbles, and boulders—wear away the rock of the sides and bed of streams.

Except for the extreme deserts and permanently frozen regions, the details of the subaerial landscapes of the world are, by and large, the result of erosion by running water. Many people find it difficult to accept the idea that great canyons like those of the Feather River system were produced by the streams that flow through them, but the truth of the concept has long been established. Erosion, given sufficient time, can reduce a great mountain system to a lowland. It has done so repeatedly, over the world, throughout geologic time.

DEPOSITION

Whenever the currents transporting sediment become slowed, sediment is deposited, either temporarily or permanently. In either case, the process of sorting or segregation occurs. Most commonly, particles are sorted according to size; pebbles are deposited in one place, sand grains in another, and clay or mud in still another, as shown in fig. 7. Because transporting currents change with time, one may expect to find that the material deposited at any given place may also change with time so that, for example, if a current is slowing down, pebbles become overlain by sand, and that in turn by mud.

Under some circumstances, mineral grains are separated according to their specific gravity. For example, gold is concentrated at the bottom of the gravel in the river beds of the Sierra Nevada because it is heavy. When recovered by panning, it is found to be associated with other heavy, though less valuable, minerals.

Ultimately all sediment comes to rest, and the sites or environments of deposition are highly varied. They include, among others, the floodplains of rivers, the sand dune areas of deserts and sea coasts, the mud flats of estuaries, and the floors of lakes and the sea.

SEDIMENTARY ROCKS

Once deposited, the sediment constitutes a body of rock or a part of a body of rock, different in larger or smaller degree from its parent, depending on its history during weathering, transportation, and deposition.

One may be disinclined to think of beach sand or mud as rock, yet hardness is not a requisite property of rock, although sediment once deposited does tend to become hardened or lithified. Lithification can come about in many ways, such as by compression and loss of water from the spaces between grains, which increases their coherence; by the deposition of mineral matter between grains, which acts as a cement; and by changes in the mineralogy of the sediment. Thus mud passes into mudstone, or, if it is very thinly layered, into shale.

Fig. 7. Sorting of sediment by currents of water. View of the Middle Fork of the Feather River looking upstream from the bridge on the Quincy–La Porte Road. Sand and gravel are separated and deposited in different places.

Sand becomes sandstone, and gravel or pebbles become conglomerate. Sediments in which the larger rock fragments escaped rounding are called breccia; but breccias, which are rocks composed of angular fragments of other rocks, are also of other origins, some of which are mentioned and explained in the text that follows. Lithification is a gradual process so that sedimentary rocks show various degrees of hardening.

Some of the products of rock weathering are soluble in water and are transported in solution in streams and in waters that pass underground. Water that contains much of such material in solution is called hard water. Some of the dissolved material may be deposited as cement between the grains of sedimentary rocks and some may be deposited at the orifices of springs, where underground water returns to the surface; an example is shown in fig. 8. Much, however, is deposited from water in which it has become concentrated by evaporation. Such deposition occurs now in desert regions and in the past it occurred in shallow arms of the sea, which resulted in extensive deposits of common salt, gypsum, and other less common minerals. Such rocks are called evaporites and are classed as sedimentary.

Much of the dissolved material resulting from weathering has another fate. For example, many organisms, such as corals and clams,

Fig. 8. Travertine, a variety of limestone deposited from underground water that has returned to the surface as a spring. Along State Highway 89 near Indian Falls on Indian Creek.

are able to extract calcium carbonate from solution and incorporate it into their body structures. Others, notably the microscopic plants called diatoms, do the same with silica. Such secreted mineral matter is left on death of the organism and may accumulate to form extensive deposits of organic sediment. Rocks composed of calcium carbonate are called limestone and those composed of diatoms are called diatomite. Chert, a common rock of the Sierra Nevada, contains the microscopic marine siliceous organisms known as radiolaria. All rocks composed of organic remains are also sedimentary.

Volcanism originates within the earth, but adds rock to the surface, both above and below sea level. Much volcanic rock is in the more or less massive form of lava flows, but much is in the fragmental form of ash and volcanic breccia. Ash is the name applied to the finest-grained product of volcanism, especially of explosions. Breccia consists of larger fragments of angular aspect that have sharp or rough edges.

If the fragmental products of volcanic eruptions become subject to transportation by water, they may quickly take on the properties of

Fig. 9. Volcanic mudflow breccia. An unsorted deposit of the volcanic rock andesite. Nonvolcanic rock may be picked up by mudflows, as shown here by the two light-colored blocks that are of coarse-grained granitic rock. Bonta Formation. By the road from Lake Davis to Mt. Ingalls, near the SE Cor., Sec. 12, T. 24 N., R. 12 E.

other sedimentary rocks except that the constituent particles may still be recognizable as having been of volcanic origin.

One must keep in mind also that volcanic rocks, like other rocks, are subject to conversion to sediment and sedimentary rock through weathering, transportation, and deposition. If the volcanic character of the rock fragments can be recognized, the resulting sedimentary rocks may be designated volcanic conglomerate, sand, or mud. Volcanic breccia may be either sedimentary or of direct volcanic origin. A distinctive rock, very abundant in this region, is volcanic mudflow breccia, which consists of angular to slightly rounded blocks of volcanic rock in a matrix of volcanic mud. An example is shown in fig. 9. Both blocks and matrix seem to have traveled from a source to the site of deposition as a mud stream without becoming sorted. The origin of volcanic mudflows, discussed further on, is still somewhat uncertain.

It should be evident even from this brief discussion that the processes that lead to the formation of sedimentary rocks may result in rocks that are quite different in composition from the parent material, and that are highly varied among themselves. The processes of weath-

Fig. 10. Well-defined beds of fine-grained sedimentary rock on the left. This is the diatomaceous and carbonaceous lower part of the Mohawk Lake Beds. The dark layers are rich in carbon from aquatic plants. In the bank of the Middle Fork of the Feather River 0.25 mile downstream from the bridge at Mohawk. SW 1/4, Sec. 9, T. 22 N., R. 12 E.

ering, transportation, and deposition tend to lower high-standing areas and fill in low-standing areas; they tend to reduce the relief of the earth surface until interrupted by other processes.

A further aspect of the sedimentary rocks that results from the processes described is that they are deposited in layers that are called beds if they are thick, or laminae if they are very thin. Such rocks are said to be bedded or layered. Bedded rocks, including the volcanic rocks, accumulate at the sites or areas of deposition as a pile or stack of layers. It is usually assumed that the layers were horizontal at the time that they were deposited, which, though not quite true, is true enough for most purposes. It is a truism that in such a stack of beds, those that are below are older than those that are above. Well-defined beds in the Mohawk Lake Beds are shown in fig. 10.

Processes internal to the earth disturb rocks at or near the surface. Rocks become deformed and the deformation is well recorded in layered or bedded rocks. The layers are often bent or folded into arches and troughs, as shown in figs. 11, 35, 43, and 44, or they are broken along surfaces called faults, as in figs. 12 and 36, in which case the rocks on one side of the fault surface have moved over those of the other side. Folding and faulting either alone or in combination can accommodate profound disturbances such that layers once horizontal and continuous are turned up to vertical, and even past the vertical so as to be overturned or upside down. For this reason it is important in studying such rocks to be able to distinguish the top from the bottom of layers. Fortunately there are features of rock, some of which are described below, that permit this distinction.

Commonly, inclined layers are present within a bed of sediment whose top and bottom are parallel. Such a layer is said to be cross-bedded (see figs. 13*a* and 55). A bed that contains cross laminae may be less than an inch thick or as much as tens of feet thick. The cross beds are usually cut off sharply at the top, but at the bottom they curve into the base, hence permitting the recognition of the top and bottom. Cross beds form by deposition from currents. Observations of ripple marks in stream beds or on tidal flats show that small cross beds result from the erosion of the gentle long slope of a ripple and deposition on the steeper, downstream slope. Large cross beds form in windblown dune sand and in coarse sediment deposited from fast-moving currents of water.

At places where sediment is being deposited currents become locally erosive and cut channels into the surface of deposition. Such channels may be less than an inch deep or as large as the channel of a large river. The sediment that subsequently fills a channel differs from the material of the walls of the channel in some way such as color or grain size, so that the filled channel can be recognized, as shown in figs. 13*b* and 56. Obviously the shape of the channel indicates the top side of the sediment that fills it.

Finally, many beds of sand are graded (see figs. 13*c*, 51, and 54); that is, the sand or gravel is coarser at the bottom of the bed than at the top, and the change is progressive upward. Graded beds are good indicators of the direction of top. A graded bed may be only a fraction of an inch or several feet thick. Graded beds, cross-bedding, and cut-and-fill channels are all present in the single block of sandstone shown in fig. 14.

GEOLOGIC TIME AND THE AGE OF ROCKS

Geologists and paleontologists have found that the life of the earth has changed with the passage of time, and that the changes are recorded in the sedimentary rocks by the preserved remnants of that life, called fossils. Even though the changes in life have not been sudden compared to man's life-span nor, in all cases, entirely worldwide, they have occurred during short enough time-spans and over large

a

Fig. 11. **a,** A small fold in a layer of chert in slate of the Shoo Fly Formation. In the bed of the Middle Fork of the Feather River downstream from the north end of the bridge on the Quincy–La Porte Road. The fold is said to be isoclinal because the sides, or limbs, are parallel. **b,** Diagrammatic explanation of **a.**

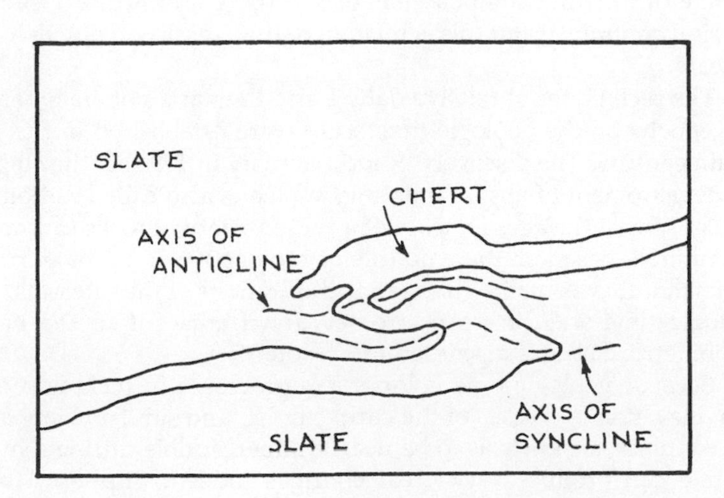

b

Fig. 12. Faults in the Mohawk Lake Beds. The nearly horizontal layers of white volcanic ash are offset across the faults. A road cut on State Highway 70 a half mile east of its junction with State Highway 89 near Blairsden.

enough regions to permit the erection of the geologic time scale shown in fig. 15, and to permit the placing of any sequence of rocks in the time scale if enough fossil remains can be found. Thus the rocks are said to be dated. They are also correlated to other sequences of sedimentary rocks with similar fossils, or with fossils of the same age, in spite of differing compositions of the rocks and in the absence of physical continuity. But this is relative dating; fossils do not yield ages in years.

The techniques of relative dating and the names of eras, periods, and epochs on the geologic time scale were established in the nineteenth century. The discovery of radioactivity in 1896 set the stage for the development of absolute dating, which is also called radiometric dating, so that the ages in years of a wide variety of rocks can now be determined, based on the amounts of radioactive and radiogenic elements that they contain. The ages and durations of the intervals of the geologic time scale in years are set down adjacent to the names. Radiometric dating is discussed in Chapter 5.

Because geologic time is long, the processes of rock decay, erosion, elevation of masses of the earth's crust, and similar changes that all take place so slowly as to be nearly imperceptible during a human lifetime can produce very great changes indeed. Suppose, for example, that somewhere the summit of a mountainous region like the

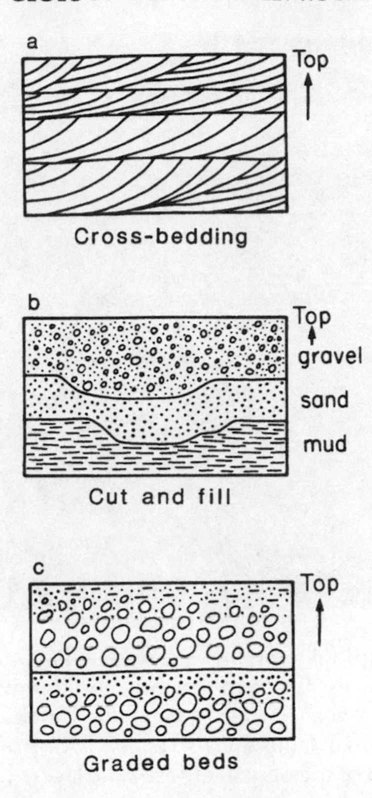

Fig. 13. Some features of sedimentary rocks that indicate the direction of accumulation—that is, that provide a distinction between the top and bottom of a set of layers. *a*, Cross-bedding. *b*, Cut and fill. *c*, Graded beds.

Sierra Nevada, elevated by forces internal to the earth, is being lowered by weathering and erosion at the rate of one thirty-second of an inch per year. In a lifetime of seventy-five years that amounts to about two and a quarter inches, hardly noticeable to anyone but a surveyor, but in a million years, a short time geologically speaking, it comes to almost half a mile.

IGNEOUS PROCESSES AND IGNEOUS ROCKS

From time to time in the earth's history, and at various places, rock well below the surface of the earth has been heated and thereby liquified to form magma. Eventually the magma cools and solidifies into the rocks called igneous. Magma tends to rise, or intrude, toward the surface, but not all arrives there. That which cools well below the surface crystallizes as a body that is said to be intrusive, whereas that

Fig. 14. Cut-and-fill structure, graded beds, and cross-bedding together in a loose block of volcanic sedimentary rock of the Peale Formation near Frazier Falls. Is the block right side up or upside down? Small faults are also present, one of which, directly below the point of the pencil, crosses the layers perpendicularly.

which reaches the surface is called extrusive. At the surface magma forms a volcanic rock, and a volcanic eruption is said to have occurred.

Only under very special circumstances, such as at the volcano Kilauea in Hawaii, can magma be sampled, so the composition of magmas is inferred from the composition of the igneous rocks that form from them. The difference is certainly not great, but magmas can contain as much as 10 percent of water and other substances that would be gaseous or liquid under conditions at the surface of the earth. Water escaping from the magma is the principal cause of volcanic explosions.

Much magma, especially that which goes to form the granitic rocks described in Chapter 5, is relatively rich in silica and poor in iron and magnesium, a fact made evident by the presence in the rocks of the glassy-looking mineral quartz and the scarcity of the dark-colored, iron-bearing minerals. Rocks from silicic magmas tend to be light colored because they are mostly of the light-colored minerals quartz and feldspar and contain only small amounts of the dark-colored minerals that are rich in iron and magnesium. At the opposite extreme are the magmas called mafic, which contain less silica and

THE GEOLOGIC TIME SCALE

Era	Period	Epoch	Duration	Age
Cenozoic	Quaternary	Holocene	0.01	0.01
		Pleistocene	1.5	2
	Tertiary	Pliocene	4	6
		Miocene	18	24
		Oligocene	13	37
		Eocene	21	58
		Paleocene	8	66
Mesozoic	Cretaceous		78	144
	Jurassic		64	208
	Triassic		37	245
Paleozoic	Permian		41	286
	Pennsylvanian		34	320
	Mississippian		40	360
	Devonian		68	408
	Silurian		30	438
	Ordovician		67	505
	Cambrian		65	570
Precambrian			3,400	4,000

The numbers are approximate times in millions of years.

Fig. 15. The geologic time scale.

correspondingly larger amounts of iron and magnesium. These yield rocks with feldspar but without quartz, and in which dark-colored minerals constitute half or more of the rock.

Because the temperature of the earth increases with depth below the surface, magma at greater depth cools more slowly than that which cools at lesser depth or on the surface. The number of crystals that begin to form in a given volume of magma depends in part upon the rate of cooling. Slowly cooling magma develops few crystals; accordingly rocks that cool slowly at great depth are coarser grained than those that cool more quickly at lesser depth. Much magma that has cooled slowly and developed some large crystals is then intruded to a shallower position, where the remaining liquid cools more rapidly, or it is extruded to the surface, where the remainder cools very rapidly indeed. In such instances many new centers of crystallization arise in the remaining liquid, causing it to form a finer-grained aggre-

Fig. 16. Porphyritic igneous rock. The large, light-colored grains called phenocrysts are embedded in a darker fine-grained matrix called groundmass. That the phenocrysts are crystals is shown by the fact that they are bounded by pairs of parallel edges that represent crystal faces in edge view. The groundmass was still liquid when the phenocrysts reached their present size. The phenocrysts that were originally feldspar and the several minerals of the groundmass were changed to other minerals when the rock was metamorphosed. Along the "C" Road 0.25 mile north of the railroad at Clio.

gate in which are embedded the fewer large grains previously formed while the magma was cooling more slowly. The resulting rock is said to be porphyritic and the larger crystals are called phenocrysts. The phenocrysts are true crystals, having formed freely in a fluid medium. That they are bounded by crystal faces is evident from the fact that their outlines seen on smooth rock surfaces are a series of straight lines that are commonly in parallel pairs. An example is shown in fig. 16.

In the extreme case of extrusion to the surface, or beneath the sea, some or all of the liquid magma may not have had time to crystallize; it quickly solidified to natural glass, the best-known form of which is obsidian.

Igneous rocks that are formed by the cooling of magma in the deepest and hottest environments—that is, several miles below the surface—are quite uniform in grain size, and the grains are large enough, an eighth to a quarter of an inch across, to be easily visible to

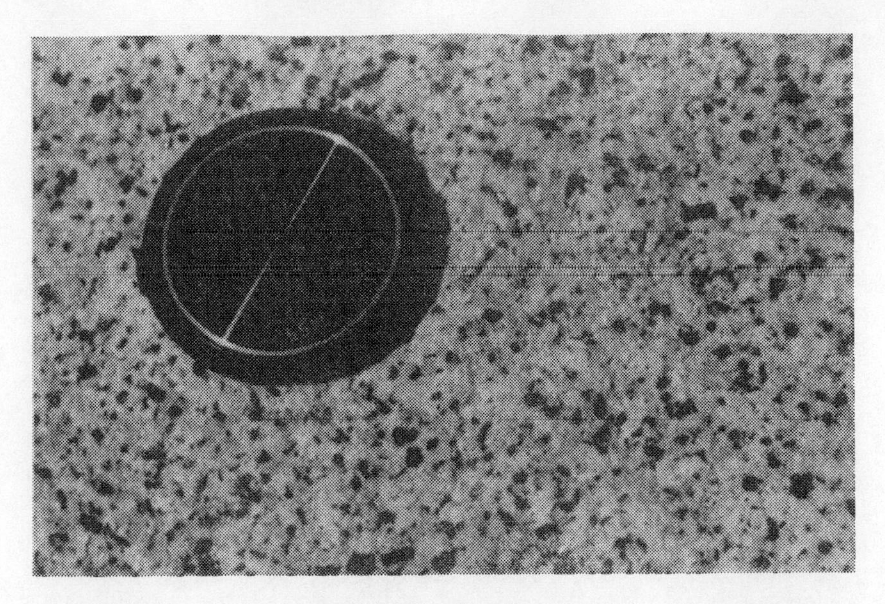

Fig. 17. Granitic rock. The grain size is slightly less than an eighth of an inch. The white grains are feldspar, and the gray grains are quartz. The black grains are both hornblende and the black mica biotite. This is the rock of the Grizzly Pluton at the rest stop on State Highway 70 near Cresta. See also fig. 76. The lens cap is 2 inches in diameter.

the unaided eye, as shown in fig. 17. Such rocks are called plutonic, a name derived from Pluto, the ancient Greek god of the subterranean world of the dead.

Many names are applied to plutonic rocks, depending on the kinds of minerals and the relative amounts of them that they contain. Most plutonic rocks in our region are silicic; they are composed mostly of quartz and one or two minerals of the feldspar family, with small amounts of one or two minerals that are black or dark green because they are rich in iron and magnesium. Since in these respects the rocks are similar to granite, it suffices to call them collectively granitic rocks. On the other hand, the mafic plutonic rock called gabbro (fig. 18), which is about half dark-colored minerals, is present near Johnsville, among the granitic rocks east and north of Plumas-Eureka State Park, and elsewhere in the northern Sierra Nevada.

Plutonic rocks with very little or no quartz or feldspar—that is, composed entirely or mainly of the dark, iron-rich minerals—are called ultramafic. Although they are much less abundant in general than are the granitic rocks, a variety called pyroxenite (see fig. 78) is present near Johnsville, and another called serpentinite is present in

Fig. 18. Gabbro. A dark-colored (mafic) plutonic igneous rock. The grain size is the same as that of the granitic rock of fig. 17, but dark-colored grains constitute about half of the rock. Eureka Peak, Plumas-Eureka State Park.

Big Jamison Canyon. Very large masses of serpentinite occur farther west in the Sierra Nevada.

Not only do igneous rocks have names, but the individual bodies or masses have names, depending on their size and shape. A general name for bodies of plutonic rock is pluton. Most plutons are quite large, have steep sides, a domed upper surface, and no visible bottom. In many instances the heat of plutons has produced some visible change or metamorphism in the older rocks into which they were intruded.

Among smaller intrusive bodies, dikes are probably the most common (see fig. 6). They result from the injection of magma into a crack in the rocks so they are thin compared to their lateral or vertical extent. Some dikes are only a fraction of an inch wide and a few feet long, whereas others are tens or even hundreds of feet wide and many miles long. Most, however, are probably from a few inches to a few feet thick, and correspondingly short.

Some dikes are granitic, being as coarse grained as the rocks of plutons, but mostly, because dikes are smaller, they cool more quickly and are finer grained. Many dikes are subject to the two stages of cooling mentioned above that result in a porphyritic rock. The early formed grains, the phenocrysts, are in some instances quite large.

At the other extreme, dike intrusions that reach the surface become the feeding vents for lava flows and are composed of rocks like the volcanic rocks described below. The rocks of dikes vary widely in composition.

Sill is a name applied to injected bodies that are tabular, as are dikes, but whose sides are parallel to the layers of rock into which they are intruded. The name is usually restricted to bodies whose original position was nearly horizontal. Like dikes, sills can be very small or enormous, and the rocks of sills also vary widely in composition.

Volcanic eruptions occur and volcanic rocks are produced when magma reaches the surface of the earth, either on land or beneath the sea. Magma reaches the surface either through a fissure or crack, or through a tubular or pipelike vent. Because eruptions from centers of volcanic activity may be long continued and because the composition of the magma erupted may change with the passage of time, the nature of the eruptions and the products may also change with time. Accordingly, what follow are generalizations to which there are many exceptions.

Mafic magma is very fluid; it often erupts quickly and quietly to form thin and extensive lava flows. However, the early part of an eruptive period may be explosive because that is the time during which the rock adjacent to the vent, originally cold, is being heated by the rising magma. Thus, the magma loses heat. In consequence, it becomes more viscous and the gases, principally water, have difficulty in escaping; they become entrapped and so explosive activity occurs. Early explosive activity is commonly followed by a period of quiet extrusion of magma.

At the other extreme is silicic magma, which is very viscous. The gases escape only with great difficulty, and most eruptions are explosive. Most of the volcanic products are fragmental; lava flows are uncommon and small. Most of the silicic volcanic rocks are porphyritic.

Between the mafic and the silicic are magmas of intermediate composition. They may behave like either the mafic or the silicic or in an intermediate manner depending on their composition, the degree of chilling to which they may be subjected, and the amount of water they contain. Long-continued eruption of intermediate magmas from single volcanic centers has resulted in the building of spectacular and beautiful giant cones, exemplified by Mt. Shasta, that are composed of both lava flows and the fragmental results of explosions. Intermediate magmas have also given rise to the most tremendous volcanic explosions of historic times. Most volcanic rocks of intermediate composition are porphyritic.

The bulk of all volcanic rocks are mafic; the silicic rocks are less abundant. The intermediate rocks are intermediate in volume as well

Fig. 19. Scoria. The volcanic rock basalt that is mostly holes. The holes were bubbles of steam escaping from the magma as it congealed. Material like this is abundant around the summit of Mt. Ingalls. The paper clip is 1.2 inches long.

as in composition, and ultramafic volcanic rocks are almost negligible in quantity.

Much magma congeals to the rock basalt, which is typically very dark gray to black, compact, and without grains visible to the naked eye. Some of the basalt of this region is like that, especially that of the Lovejoy Formation, but that of the Warner Formation is light to dark gray, and usually porphyritic with light green to golden-yellow phenocrysts of the mineral olivine. Phenocrysts of white feldspar or green pyroxene are also present in some lava flows of the Warner Formation.

Basalt commonly contains spherical or oval-shaped holes, called vesicles, that are caused by escaping steam while the rock is hardening. Each hole is a vesicle and such rock is said to be vesicular. Rock that has so many holes as to constitute a rock foam is called scoria (fig. 19) and is said to be scoriaceous. In some places the holes have become filled with mineral matter: the fillings are called amygdules, their shape being commonly that of the almond (see figs. 59 and 67). Such rock is said to be amygdaloidal. Commonly the upper and lower parts of lava flows of basalt are vesicular or scoriaceous and the central part is massive, that is, without holes.

Fig. 20. Fans of poorly formed columns in the black basalt lava of the Lovejoy Formation. Along Red Clover Creek. NE 1/4, Sec. 25, T. 25 N., R. 12 E. See also figs. 6 and 125.

Lava flows of basalt are usually jointed or cracked owing to the cooling process. Best known perhaps are the columnar joints that divide the rock into postlike columns, often fairly regularly hexagonal in cross section. The columns extend through the lava flow about perpendicular to the upper and lower surfaces. However, they are commonly imperfectly formed and in some places fans of columns occur, as in fig. 20. Other joints are parallel to the upper and lower surfaces of flows and divide the rock into plates that make good flagstones.

All of these features can be seen in the basalt lavas of the Warner Formation on Penman Peak, Mt. Jackson, and Mt. Ingalls.

Explosive eruptions of basalt produce scoriaceous fragments that accumulate around the vents as scoria cones, also called cinder cones. Some scoria cones are brick red rather than black, an effect produced by escaping steam. Part of a cinder cone enveloped in lava of the Warner Formation on Mt. Ingalls is shown in fig. 21.

Basalt magma extruded underwater may be chilled very rapidly, especially in the deep sea, where the water is very cold. Sometimes the result is a deposit of fragments of glass, but angular fragments of mostly crystalline rock are also common. Perhaps submarine volcanoes are largely great heaps of fragmental rock. However, under

Fig. 21. Bedded basalt ash. Part of a cinder cone once buried by lava of the Warner Formation and now exposed by erosion. On the north slope of Mt. Ingalls. NW 1/4, Sec. 27, T. 25 N., R. 12 E.

some circumstances the congealing magma does not fragment; it develops a glassy skin that expands into the shape of a pillow or a well-filled sack. The pillows break off and new pillows form so as to produce a heap of pillow-shaped masses, as seen in figs. 22, 23, 59, 61, and 62. The original skin is glassy and the interior is usually crystalline and fine grained, often amygdaloidal. The result is pillow lava or pillow structure, which is definitely diagnostic of subaqueous extrusion. At some places, highly deformed sediment is present between the pillows, as shown in fig. 23.

Magmas of intermediate composition give rise to a variety of rocks whose names need not concern us. The intermediate rocks of this region are mostly the usually porphyritic rock known as andesite. Andesites are varied in color, being in shades of gray and brown, many with a pinkish cast; some verge on rusty red. Most contain phenocrysts of white or gray feldspar and one or more dark green or black minerals. Many of ours contain phenocrysts of the distinctive and easily recognized shiny dark brown or black mineral hornblende (fig. 24). Much andesite is vesicular but the cavities are often small and irregularly shaped. The phenocrysts of some andesite lavas are oriented in a common plane or direction by flow (fig. 25), and some rocks have a layered structure of thin alternating layers of different colors, also caused by flow of the magma (fig. 26).

Fig. 22. Pillow lava of the Taylor Formation. The cross sections of pillows exposed on a glaciated surface. The dark rims were originally glass. The highly deformed material between the pillows was layered andesitic mud into which the pillows rolled when they became detached from their source. The handle of the hammer just above the center of the picture is 14 inches long, and the largest diameter of the pillows is about 4 feet. Located about 1,000 feet west of the E 1/4 Cor., Sec. 6, T. 21 N., R. 12 E. It is east of Gray Eagle Creek and about 1,100 feet west of the Gold Lake Highway.

Most intermediate magma is erupted through central tubular vents, which are filled with igneous rock when activity is finished. Removal of superficial material by erosion often reveals a massive body called a plug or neck, commonly more resistant to erosion than the surrounding rock. An excellent example is Sugar Loaf on the north side of the Sierra Valley 3 miles northeast of Beckwourth, shown in fig. 27. However, many plugs in the northern Sierra Nevada and Diamond Mountains are composed of a breccia of angular fragments of volcanic rock, two of which are illustrated in figs. 114–119.

Silicic magmas are prone to erupt explosively, hence to give rise to deposits of fragmental material. The magma seems to rise to near the surface, where it becomes very viscous. Escaping water forms vesicles that, owing to flow movements, are often drawn out into subparallel tubes. The magma then hardens to a porous mass of natural glass that resembles taffy candy. The resulting rock, called pumice (fig. 28), is light colored and much of it is so lightweight that it will

Fig. 23. Part of fig. 22 in detail, showing the highly deformed thin-bedded volcanic (andesite) mud between the pillows.

Fig. 24. The intermediate volcanic rock andesite. It is porphyritic, with phenocrysts of shiny dark brown to black hornblende. That many of the phenocrysts are elongate in a common direction is the result of flow of the magma. The large white spot is enamel. The paper clip is 1.2 inches long. The specimen is from the Penman Formation.

Fig. 25. The mafic volcanic rock basalt with abundant phenocrysts of light-colored feldspar. The book-shaped feldspar crystals are arranged with their broad surfaces in a common plane, a consequence of flow of the magma. The paper clip is 2 inches long. The sample is from the Lovejoy Formation at the head of Cogswell Ravine above the forest road from the Jackson Creek Campground to Happy Valley. Pieces of the rock can be found along the road in Sec. 17, T. 23 N., R. 12 E.

float on water. If the gas pressure in the congealing magma increases until it exceeds that of the overlying material, an explosion occurs. Fragments of pumice and pulverized pumice are expelled, and cones of light-colored pumice similar in form to the scoria or cinder cones of mafic rocks are the result. Viscous magma may well upward into the crater of a cone to form a bulbous protrusion called a plug dome. Thick lava flows of small extent may also form. Many such domes and lava flows are of the compact natural glass obsidian. Cones of pumice, plug domes, and lava flows of obsidian comprise the Mono Craters and can also be seen near Medicine Lake east of Mt. Shasta. None are known in the Feather River country.

Pulverized pumice and fragments of pumice are lightweight, and light colored. The fine particles, often widely distributed by wind and running water, give rise to deposits called ash whether the final resting place is above or beneath water. Ash is not, of course, the re-

Fig. 26. Andesite from the bottom layer of the Penman Formation on Grizzly Ridge. The varicolored layers are the consequence of the magma flow. As a result of falling temperature the viscosity of the magma increased to such an extent that the flow layers became folded and faulted. The paper clip is 2 inches long. This is a polished slab.

sult of the burning of anything. Ash consolidated into a firm rock is called tuff.

Some eruptions of silicic magma that consist of very large volumes of pumice charged with very hot gas, mostly water vapor, result in pyroclastic flows like that erupted at Mt. St. Helens in May 1980. Pyroclastic flows are discussed in more detail in Chapter 7.

ROCK METAMORPHISM AND METAMORPHIC ROCKS

Metamorphic rocks are sedimentary or igneous rocks that have been changed by high temperature and pressure and the chemical activity of solutions. Metamorphic rocks can also be metamorphosed repeatedly.

The essential process in metamorphism is recrystallization, which means simply that changed conditions of temperature and pressure cause the original minerals of the rock to change to different minerals. Most metamorphic rock is characterized by a distinctive feature called

Fig. 27. Sugar Loaf. A volcanic plug or neck of andesite. It is the congealed magma that filled the throat of a volcanic vent. The present erosion surface is far below any surface upon which the volcano may have erupted. North of State Highway 70, 3 miles northeast of Beckwourth.

Fig. 28. Pumice. A froth of silicic lava from the eruption of Mt. St. Helens, Washington, May 1980. The paper clip is 2 inches long.

cleavage, which is not present in unmetamorphosed rock. Cleavage is the property that permits rock like slate to split along closely spaced planes. Metamorphism is a complex subject and not everything about it is yet entirely clear. How it comes about can best be explained in terms of plate tectonics, a concept already mentioned at the beginning of this chapter.

Crustal plates separate and "grow" at the midocean ridges by the intrusion of basic igneous rock (mostly basalt) derived from the underlying mantle. Because in this way the plates are enlarged, at some other place they must be reduced. At some edges plates slip past each other, but at most edges an active plate moves down and under an opposing plate along a subduction zone, as shown in fig. 3. Subduction zones are great inclined faults whose surface expression is a very deep trench in the sea floor. They are the sites of many great earthquakes and the locus of formation of magma. The leading edge of the active plate is said to be subducted. As it is subducted it becomes warmer and eventually is reincorporated into the mantle. Subduction occurs at the boundary between two plates of oceanic crust, or one of oceanic and one of continental crust.

During their long, slow travel, oceanic plates become coated with a veneer of volcanic rock, mud, and organic sediment. Where an opposed plate is continental rock, as is commonly the case, much sediment derived from the continent is also shed into the sea at the plate margin. As the active plate is subducted, its veneer of volcanic rock and sediment seems to be scraped off, as it were, and piled up against the opposed plate. There it becomes mixed with sediment from the opposed plate and with new volcanic rock generated by the subduction process. At times during this complex process, parts of the ocean crust igneous rocks and even part of the mantle become mixed with the accumulating sedimentary and volcanic rocks.

The result is a long, narrow belt of deformed sedimentary and volcanic rock called a geosyncline (see figs. 39 and 72*B–B'*). The word geosyncline originally meant a very large trough or downfold in the earth's crust located along a continental margin. It was thought that the trough became filled with sedimentary rock, which later became folded into mountains. That notion of the geosyncline is superseded by the concept of plate tectonics, but as yet there is no better word to describe the accumulated sedimentary and volcanic rock that is now believed to be associated with subduction. So the word is continued in use but in a new sense, in reference especially to the accumulated rock, and not to the configuration of the floor beneath it.

Geosynclines occur not only along continental margins, but also in ocean basin associated with volcanic island arcs, like the Aleutian Island chain, where opposing plates collide.

As subduction proceeds, the accumulating geosynclinal rock, which may become many miles thick, is also carried down to a region of higher temperature, where it becomes heated. It is also intensely deformed in the vise between the opposed and moving plates and so it becomes intricately folded and faulted. These are the conditions for metamorphism. New minerals appear because of increased temperature and pressure and the presence of solutions that were perhaps mostly sea water entrapped in the pore spaces between the particles of sediment and fragmented volcanic rocks. This is recrystallization. Because of the mechanical effects associated with the deformation, new minerals become oriented so as to produce rock cleavage, which is described below. At appropriate depth, some of the geosynclinal rock is melted and the magma of intermediate composition rises to produce volcanoes on the opposed plate. Silicic magma forms at great depth and intrudes upward into the deformed and metamorphosed geosynclinal rocks to form plutons (see figs. 3 and 72*D–D′*). The end result is the transformation of the geosynclinal rocks from a weak mass of particles saturated with water into a body of relatively hard and rigid rock welded to the continent. In this way continents become enlarged. These processes are described in more detail in Chapters 3, 4, and 5.

What is described above is called orogeny, the mountain-building process. The result is the conversion of the submarine geosyncline into an elongate range of mountains characterized by folded and faulted metamorphic rock invaded by plutons of granitic rock.

The diversity of sedimentary and igneous rocks subject to metamorphism gives rise to an equal diversity of metamorphic rocks, complicated by variations in the intensity of the metamorphic processes, which are essentially a function of temperature. At the highest temperatures all features characteristic of the original rock are often lost; indeed, melting may be thought of as the ultimate stage of metamorphism. At lower temperatures, however, many features of the original rock are preserved, and that is true of most of the metamorphic rock of our region.

Most of the metamorphic rock of the Sierra Nevada has cleavage, the most obvious feature of most metamorphic rocks. It is the property that permits the rock to be split on very closely spaced, rather smooth surfaces that, in some instances, are nearly perfect planes. It is exemplified in the familiar rock slate (fig. 29). Slate can be cleaved because it is composed in large part of a multitude of flakes of mica or micalike minerals. Each grain of the micaceous mineral has a single planar direction of mineral cleavage, or easy splitting; superficially it resembles a tiny tablet of paper. All the grains are arranged so that their directions of cleavage are in a common direction. Thus the abil-

Fig. 29. Slate with perfect cleavage. In the Shoo Fly Formation. Along the Johnsville–La Porte Road at the Plumas-Eureka State Park boundary. SW 1/4, Sec. 26, T. 22 N., R. 11 E.

ity of the rock to be split is a reflection of the ability of the mineral grains to split.[1]

The cleavage surface is commonly independent of other planar features of the rock, such as bedding or the flow layers of volcanic rocks. Cleavage may be parallel to bedding surfaces but very commonly it makes an angle, often a large one, with the bedding, as shown in fig. 30. Thus cleavage is often independent of the folds in bedding, but it in turn may be folded.

The cleavage surfaces in rocks that were originally sandstone or that contained pebbles or angular rock fragments may not be planar,

1. It is also true that in some rocks that have cleavage, elongate minerals are arranged not only so that their long dimension is in the cleavage plane, but so that their elongation is in a common direction within the cleavage plane. Furthermore, it has long been known that the grains of quartz, a mineral that has neither easy cleavage nor pronounced elongation, that are present in rocks with cleavage are also arranged in a common orientation. For this reason, it is clear that the cleavage of metamorphic rocks is not merely a function of the shapes of mineral grains or of the presence of grains with mineral cleavage, but is a more fundamental property dependent on the properties of the crystal lattices, that is, the way in which atoms are arranged within the crystalline grains that make up the rock.

Fig. 30. Slaty cleavage, nearly vertical, crossing bedding that is steeply inclined to the right. The angle between them is 30 degrees. Metamorphosed volcanic ash of the Peale Formation near the picnic grounds at Frazier Falls.

as they so often are in slate, but may show on edge as sheetlike trains of mica that wind between the larger grains.

In the coarser rocks such as volcanic breccia the cleavage may in some instances be present both in blocks and in the matrix, but in other instances it may be visible in only the originally finer-grained matrix between blocks, as shown in fig. 31.

The effect of deformation during metamorphism is exemplified by the fragments of the volcanic breccia shown in fig. 32, which have been compressed into the shape of lenses. Figure 33 shows a dike that was once straight and of uniform width. Its present bent appearance is owing to slip along cleavage surfaces, whose direction is indicated by the pencil. The present width of the dike is everywhere the same, measured in the direction of the cleavage. Each cleavage surface has behaved like a minute fault. This effect can be produced by slipping the cards of a deck on the side of which a stripe has been drawn.

The nomenclature of metamorphic rocks is complex but only a few names are needed for our region. Most of the rocks were metamorphosed at such low intensity that their original character is quite clear. Such rocks may be named by adding the prefix "meta" to the name of the original rock—for example, metasandstone. The name

Fig. 31. Cleavage deviated around the blocks of a volcanic breccia. Although the blocks are very little deformed, some have "tails," as does the upper side of the block to the right of the pencil tip. Taylor Formation. Near the E 1/4 Cor., Sec. 6, T. 21 N., R. 12 E.

slate is an important exception to this rule. Slate is familiar to most people, and the name is in the common language. Furthermore, no one would ever call slate by the unlovely epithet metamudstone, which it really is.

Slate is an easily cleavable, often perfectly cleavable, fine-grained rock. It is commonly black or gray, but some is almost white and some is rusty red or purplish in color. When weathered it is often scaly or silvery in appearance. The grains of micaceous minerals that permit the cleavability cannot be seen without the high magnification of a microscope, but their presence is indicated by the fact that a freshly broken cleavage surface is more lustrous than is a surface freshly broken across the cleavage. Sandstone is commonly interlayered with mudstone, so slate and metasandstone commonly occur together. Rock that contains pebbles becomes metaconglomerate, and the presence of pebbles should be easily recognized even if they have been deformed during the metamorphism.

Metamorphosed igneous rocks are easily recognized because, in many instances, phenocrysts, amygdules, flow layers, and pillow structure are well preserved.

Silicic volcanic rocks, mostly ash and breccia of the composition

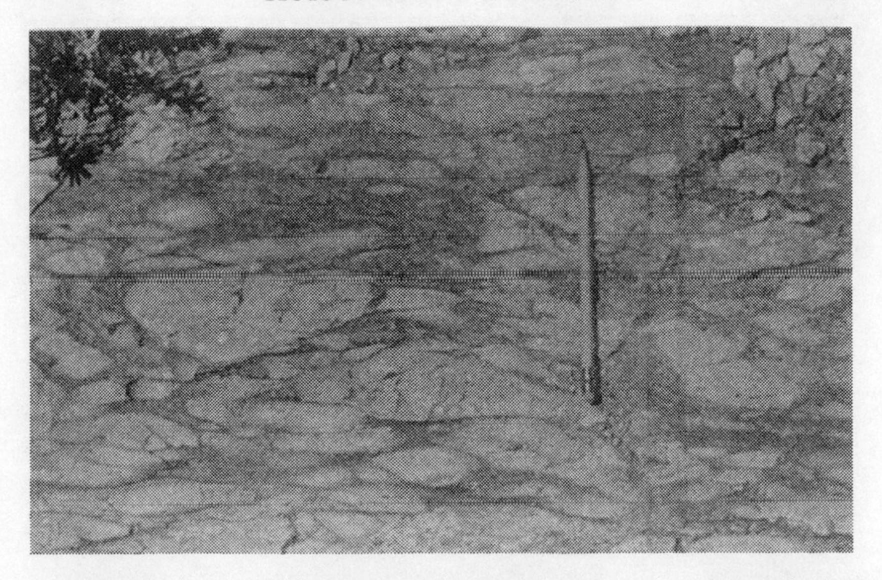

Fig. 32. Cleavage passing through the blocks of a deformed volcanic breccia. Blocks that were no doubt irregular and probably angular in form have been deformed into the shape of a lens. The cleavage direction is perpendicular to the pencil. Shoo Fly Formation, beside the Johnsville–La Porte Road, SE 1/4, Sec. 27, T. 22 N., R. 11 E.

of rhyolite are called metarhyolite or metarhyolite ash or metarhyolite breccia. Flow layering in blocks of breccia and quartz phenocrysts are commonly preserved. Even impressions of the delicate structure of pumice have been preserved in some places. Cleavage is often deviated around blocks or crystals of quartz, especially if the matrix was fine ash. Very fine rhyolite ash without crystals yields a pale greenish gray or nearly white slate.

Igneous rocks of intermediate or mafic composition, whether intrusions, lava flows, breccia, or ash, become green because the metamorphism produces an abundance of the green minerals amphibole, chlorite, and epidote. The colors of the rock may be light gray-green or various shades of green to black. They are often lumped together as greenstone, but if their original nature is clear, they are better called by such names as meta-andesite ash, metabasalt breccia, or metavolcanic sandstone.

Limestone and dolomite could be called metalimestone and metadolomite, or either may be called marble, although that common name is so indefinite in meaning that its use should be discouraged. Chert becomes metachert.

The name schist is applied to rocks that have been metamor-

a

Fig. 33. **a,** A "shear fold" in a dike. The apparent fold is not the result of bending, as is the fold shown in fig. 11, but is the result of slip on the cleavage surfaces, which acted like faults with minute offsets. Thus the distance across the dike is everywhere the same when measured along the trace of the cleavage. This is not obviously true in the photograph because the surface of exposure is irregular. This kind of folding can be produced by drawing a band on the edge of a deck of cards, then slipping the cards. In the Shoo Fly Formation at the same location as fig. 32. The pencil at the center of the picture lies parallel to the cleavage and provides a scale. **b,** Diagrammatic explanation of **a.**

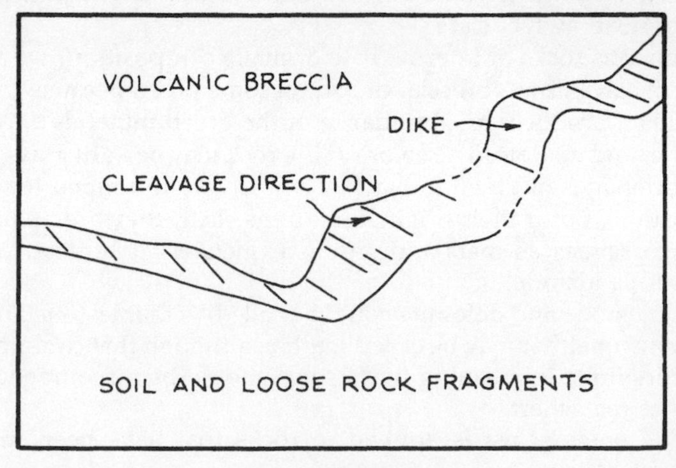

b

Fig. 34. Schist. A metamorphic rock with cleavage and a grain size like fine sand. North of Portola on the road to Lake Davis. Close to the cattle guard near the SE Cor., Sec. 22, T. 23 N., R. 13 E.

phosed at higher temperature and pressure than those described above. Schist is characterized by cleavage and by grain size about like that of sandstone. Original features of the rock are not so well preserved, or they are absent. Although there is very little schist in our region, that shown in fig. 34 can be readily visited at its place of occurrence along the road from Portola to Lake Davis.

Because magmas transport heat from their place of origin to lesser depths, the rocks adjacent to intrusions of igneous rock may have been heated to a higher temperature than those farther away. This circumstance gives rise to what are called contact metamorphic rocks. Many contact metamorphic rocks are granular and hard; many of them that once had cleavage have lost it because of recrystallization under the subsequent high temperature. Such rocks are commonly found close to the borders or contacts of plutons, but are also found at the margins of volcanic plugs where rocks have in some places been fused. Unusual minerals may be present because of the influence of exudations from the magma, or from unusually high temperature.

Most of our rocks have been metamorphosed lightly so that those features that are diagnostic of their original condition are well preserved. Because the use of the prefix "meta" results in some quite awkward words and would be endlessly repeated, it is omitted from the descriptions in the following chapters where no confusion would result from the omission. The rock names used are those of the premetamorphic parent.

NAMES OF ROCKS AND MASSES OF ROCK

Geologists use two sets of rock names. The varietal set includes most of the names used thus far in this discussion and can be applied with few exceptions to pieces of rock as small as can be held in the hand. A second set is applied to large bodies or masses of rocks, as explained below.

Rock names in the varietal set are defined mostly in terms of the minerals or other constituent parts of the rock, and the way in which the constituent parts are shaped and arranged. The chemical composition and the place of origin also enter into the schemes of classification. Some common rock names of this kind, familiar to many people, are sandstone, shale, and limestone among the sedimentary rocks; basalt (volcanic) and granite (plutonic) among the igneous rocks; and marble and slate among the metamorphic rocks. The varietal names are universal in application.

Wherever a rock as defined in the varietal set occurs, there is a body of it of finite size and, in some cases, of characteristic form. Some such bodies have already been defined: pluton, dike, plug, sill, lava flow, bed. Local geographic names are sometimes applied to such bodies, and the form of the body may also enter into the name. An example is the great mass of granitic rock crossed by the North Fork of the Feather River between Pulga and Storrie, which is called the Grizzly Pluton after one of the several Grizzly creeks in Plumas County.

There is also a need for names of assemblages or groups of layered rocks of such size that they can be shown on a map. Such assemblages are called formations, and are also given local geographic names.

A formation may be either all of one kind of varietal rock or a mixture, such as alternating layers of sandstone and mudstone, or alternating layers of lava flows and volcanic breccia, or alternations of lava flows and sedimentary rocks, and so on, which, taken together, permit it to be distinguished from adjacent formations. Formations are also characterized by having lateral extent, thickness measured perpendicular to the layers, and geologic age.

Because metamorphic rocks are also layered, they too may be given formation names if they are not so intensely disturbed as to make it impractical to do so. The Sierra Buttes Formation is composed almost entirely of metamorphosed silicic volcanic ash and breccia and is named for the Sierra Buttes. The Lovejoy Formation consists of basalt lava flows. Although it occurs in isolated patches across the Sierra Nevada and Diamond Mountains and also under the Sacramento Valley, it was obviously a once continuous sheet. It is named for Lovejoy Creek near the now inactive Walker Mine. The Mohawk Lake Beds,[2] which include all the sedimentary rock that nearly filled the basin occupied by the now extinct Mohawk Lake, are named for the Mohawk Valley.

FOLDS AND FAULTS

I have already mentioned that folds, faults, and the cleavage of metamorphic rocks are the result of deformation of the rocks. Many joints that are cracks in the rocks are also the result of deformation. Most such features, collectively called structures, are the result of forces internal to the earth; for example, the drift of crustal plates that results in subduction causes folds, faults, and metamorphism at continental margins. However, not all folds and faults are the result of subduction for some are formed well within the continental plates, far from subduction zones. Furthermore, not all folds, faults, and joints are the result of forces that originate within the earth. For example, when a landslide occurs, moved by the force of gravity, the rock of the sliding mass may be folded and faulted. Also, flowing magma as it congeals becomes viscous enough to flow in layers that are often visible in the cooled igneous rock, and it is common to find them intricately folded and faulted, as shown in fig. 26. Such small structures often mimic rather closely the large structures of folded mountains. The quite familiar columns in lava flows are separated by joints that result from forces generated by the cooling of the rock.

When force is applied to rock it may either bend or break or do both. If the rock is layered so that the effect of bending can be seen, as in figs. 11 and 43, we say that the rock has folded. If the rock breaks it may shatter into a mass of irregular fragments, or if it breaks along a single surface or a set of related surfaces along which motion occurs, it is said to be faulted.

It may be a novel idea to some that rocks can bend, because we find them so brittle when we handle them or use them as construction materials, but under the proper conditions they do bend. Under sufficiently confined conditions they flow plastically.

Geologists have a quite complex system of names for folds and faults, and whereas most of them need not concern us here, a few names and definitions are needed.

An arch in the layers of rock is an anticline and a trough is a syn-

2. The word Beds as used here is a synonym for Formation. A similar usage of a varietal rock name such as Basalt or Sandstone is also permissible.

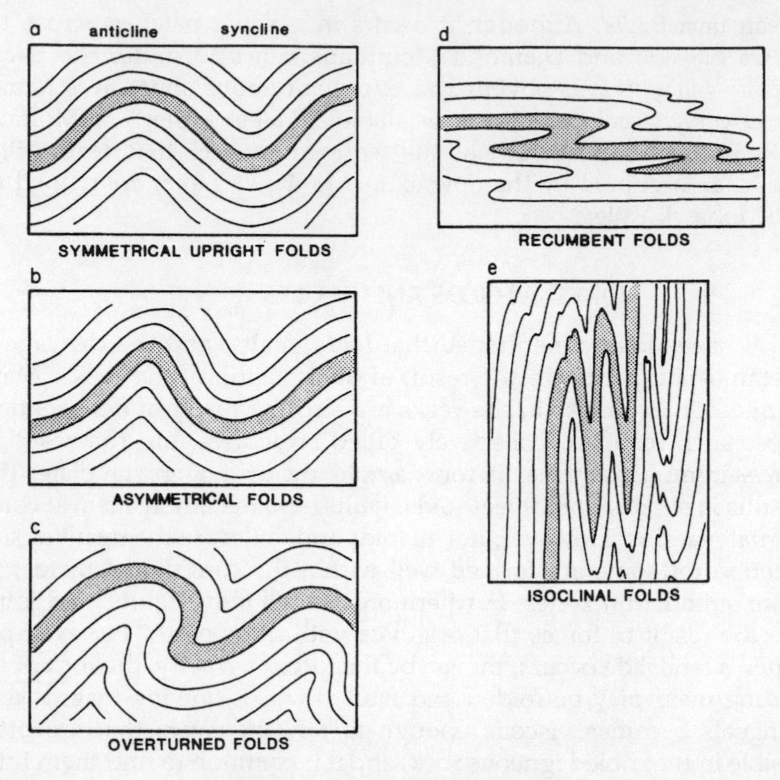

Fig. 35. Several kinds of folds.

cline. They are usually next to each other for folds come in a series like waves on the surface of water, as shown in fig. 35*a*. Those are upright folds, but not all folds are upright. Some are asymmetrical (fig. 35*b*), or overturned (fig. 35*c*), or recumbent (fig. 35*d*). Folds whose limbs are closed like the bellows of an accordion are called isoclinal, as in figs. 35*e*, 11, and 44. Because folded rocks can be refolded, and in some places in metamorphic rock one cannot distinguish the tops from the bottoms of layers, adjacent folds are often called antiforms and synforms.

Folds may be very large indeed, with a span of several miles and a length of tens of miles. Others are so small that a magnifier may be needed to examine them.

A fault is created when rocks break and the rock on one side of the break slides over the other side. Often the broken ends of distinctive layers can be identified and used to measure the amount of slip that occurred.

The vibrations (elastic waves) that are created when a fault slips are an earthquake, but not all fault movements cause earthquakes; some faults move slowly and more or less continuously without causing earthquakes, a process called creep.

Pulverized rock, called gouge, that is the result of fault movement is usually present as a slick material in the fault; it probably acts as a lubricant and facilitates creep. Adjacent to many faults the rock is fragmented into fault breccia.

Judging from historic fault movements that have broken the earth's surface, mostly at the times of large earthquakes, single fault movements are not large. Perhaps most are only a few feet, others are known to be as much as 20 feet, and one is believed to have been almost 50 feet. The measured movements of many thousands of feet to tens of miles on many faults are therefore inferred to be the sum of a great many small movements.

Different fault surfaces have different angles of inclination with respect to the horizontal. Some are very nearly vertical, as in fig. 36*a*, and are called vertical faults. At the other extreme are the thrust faults, whose surfaces may be only slightly inclined to the horizontal, as shown in fig. 36*b*. Intermediate to these are faults whose surfaces are variously inclined. In the case of thrust faults and others that are not vertical, one side of the fault is above, or over, the other. From this fact we define two more classes of faults, normal and reverse.

Normal faults are those along which the upper side has moved down with respect to the lower side, as shown in fig. 36*c*. Reverse is the opposite, the upper side having moved up with respect to the lower side, as in fig. 36*d*. Incidentally, normal is just a name—there is nothing more normal about a normal fault than there is about any other fault. Figure 36*c* also shows that a normal fault movement involves an extension of the earth's crust; that is, the upper and lower blocks move away from each other. On the other hand, in reverse and thrust faults, the upper and lower blocks move toward each other; the earth's crust is shortened. This is often called compression.

Folds also indicate that the folded rock was shortened, and although not all folds are the result of compression, one is generally correct in associating reverse faults and thrust faults with folds, and with large-scale compressions in the earth's crust. Likewise we associate large-scale normal faulting such as that in the Basin and Range Province, described in Chapter 2, with extension of the earth's crust.

One must keep in mind that examples of these various kinds of faults may be very large and significant features of the earth, or they may be very tiny and, taken singly, not very important.

Geologists have named many other categories of faults based on geometrical relationships in three dimensions, but we need to know about only one more—the lateral fault, also called the strike-slip fault,

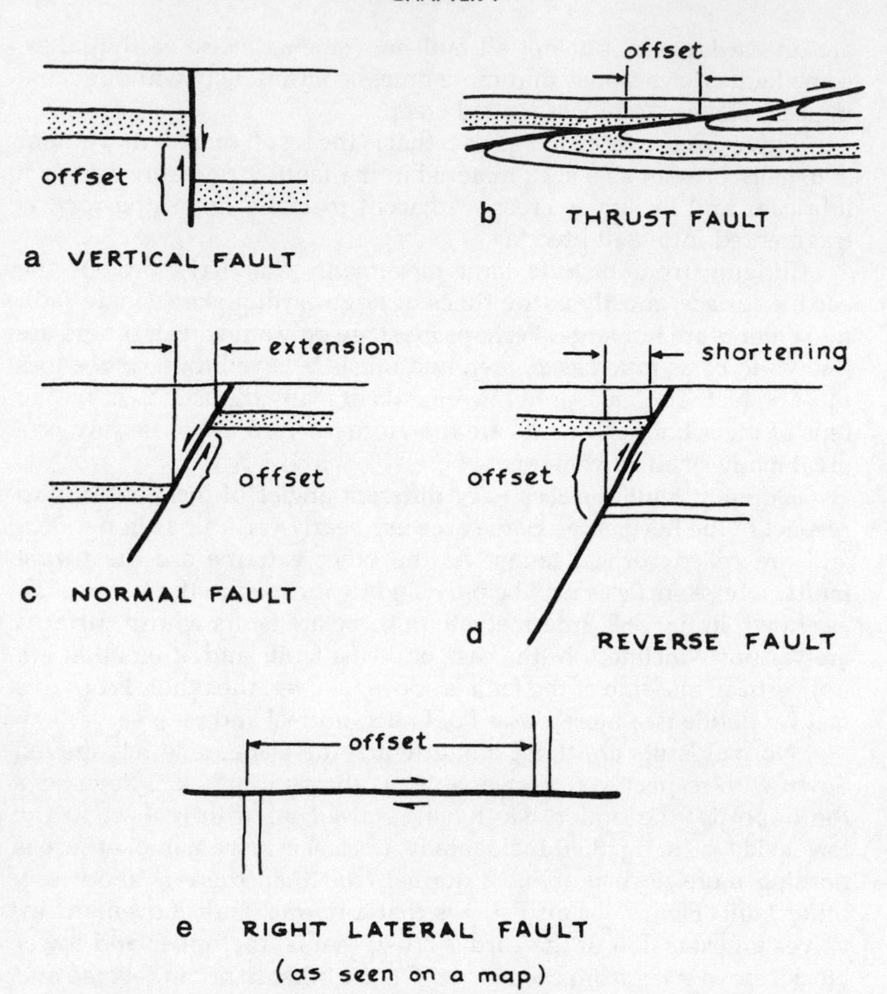

Fig. 36. Types of faults. **a–d,** Cross sections such as may be seen in a cliff or a road cut. **e,** As seen on a map. These diagrams are two dimensional. Real faults exist in three dimensions and their true relationships are much more complex than are shown here. The arrows indicate the direction of relative motion.

in which one side moves horizontally past the other, as shown in fig. 36*e*. Lateral faults are either right lateral or left lateral, and the distinction is easy to keep in mind. The observer stands beside the fault and looks across to the other side. If the other side has moved to the observer's right, the fault is right lateral, and vice versa. It makes no difference on which side of the fault one stands, the relationship is the same. The notorious San Andreas Fault of the coastal region of California, perhaps the world's best-known fault, is right lateral, so that in the course of sufficient time, San Francisco may wish to annex Los Angeles or vice versa.

Fault surfaces are not geometrical planes and some are quite strongly curved. The angle of inclination of a fault is not necessarily everywhere the same; in fact, depending on the relationship of a fault surface to folded layers, a fault may appear to be normal at one place and reverse in another.

It is also true that the sense of motion in a fault surface may change after it has once moved; that is, a fault may for a time be normal and then become reverse, or vice versa. Probably the time interval between such reversals may be either long or short.

Here we end our introduction to physical geology pertinent to the geologic history of the Feather River country. The history of that region begins, as is proper, after Chapter 2, which contains a brief definition and description of the Sierra Nevada in both geographic and geologic terms. The Feather River country is, of course, the northern part of that great range of mountains.

2

The Sierra Nevada: Defined and Briefly Described

Geologists and nongeologists differ somewhat in their views about the boundaries of the Sierra Nevada. Their differences are not great, but they are pertinent to a discussion of the geology.

To local residents, vacationers, and tourists, the Sierra Nevada includes all the beautiful forested and well-watered mountains of eastern California. It is a land of alpine landscapes, of deep canyons with fast rivers, of lakes both natural and manmade; a land for hunting, fishing, skiing, and other recreational activities. It is also a land productive of timber, minerals, cattle, and other agricultural products, and of hydro-electric power and domestic water. This is a geographical point of view, and nongeologists know perfectly well when they are in the Sierra Nevada, or in the Great Valley to the west, or in the desert to the east—the geological details are not important to them.

Geologists, on the other hand, although equally aware of esthetic and economic values, must necessarily be concerned with details of boundaries. Their most simple geologic definition is that the Sierra Nevada (fig. 37) is a block of the earth's crust about 400 miles long and as much as 90 miles wide that has been uplifted and tilted westward along a major fault system that marks its eastern limit. The process of uplift began long ago but became rapid only about 2 million years ago, during which time the present landscape took form. The uplift still continues. The geologic history of the block differs in some degree from that of the lands that surround it.

No problem exists about the boundary on the west, because not even a geologist is much concerned about standing with one foot in the Great Valley and the other in the mountains. North of the canyon of the San Joaquin River the position of the west base of the mountains is quite indefinite; the mountains rise almost imperceptibly from

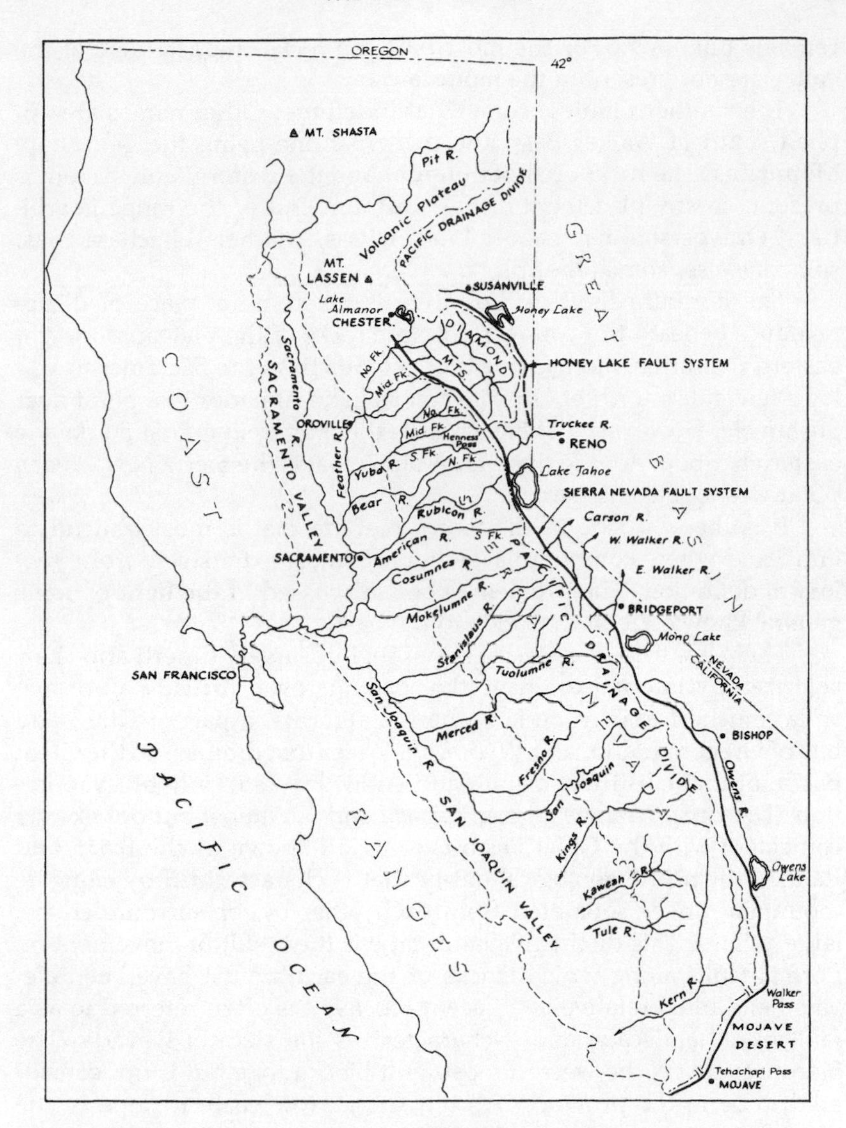

Fig. 37. The Sierra Nevada and the Diamond Mountains, the fault systems that bound them on their eastern sides, and the principal streams.

the valley as a long gentle slope interrupted only by the deep canyons of the large rivers. Toward the west the rocks typical of the Sierra Nevada continue beneath the valley floor. But south of the Kings River the west side of the Sierra is different; the mountains rise abruptly from the valley, and the reason for the difference—north and south—

remains unknown. For the most part the rocks characteristic of the valley are not present in the mountains.

The southern limit is somewhat indefinite. Other names are applied south of Walker Pass, the principal one being the Tehachapi Mountains. The rocks are not different and the eastern fault system is present, so any placement of the southern end of the range is arbitrary. One person may choose Walker Pass, another Tehachapi Pass, someone else, some other place.

The northern limit of the range is taken as its place of disappearance beneath the younger volcanic rocks of the plateau of northeastern California, along a line that extends from the Sacramento Valley a few miles north of Oroville across Lake Almanor to a point near Susanville. However, that line crosses the fault system that marks the eastern boundary of the tilted fault block that is the Sierra Nevada at a point west of Lake Almanor.

It is the east side of the Sierra Nevada that is most difficult to limit, and where common usage departs most extensively from geological definition. The problem is best discussed in the light of some general knowledge of the lands to the east.

There lies the Great Basin, a vast region, largely desert, and characterized by internal drainage that extends over virtually all of Nevada, much of eastern and southern California, a part of Utah, and bits of Oregon, Idaho, and Wyoming. Water that drains into the Great Basin, or that falls there as rain and snow, leaves it only by evaporation. The Great Basin is a geographical entity. Almost but not exactly coincident with the Great Basin is a region known as the Basin and Range Province, a geological entity that is characterized by elongate mountain ranges separated from each other by, or surrounded by, large valleys. This distinctive landscape is the result of movement on normal faults along which blocks of the earth's crust have been elevated and tilted relative to adjacent blocks. It is often referred to as a fault block landscape and it characterizes the state of Nevada. The Sierra Nevada is the westernmost fault block range but is not considered to be in the province. All but the narrow eastern slope of the Sierra Nevada is outside the Great Basin; the range may be thought of as being the western edge of the Great Basin.

It is easy to define the eastern limit along the southern part of the range. From Mohave almost to Mono Lake it is the base of the steep mountain front that is so sharply separated from the adjacent lowlands. Near Independence it is the west wall of the Owens Valley, which rises 10,000 feet in a horizontal distance of only 6 miles. The base of this front, or scarp, is marked not by a single linear fault; rather it is a zig-zag line of faults in a complex system of faults.

North of the Owens Valley the scarp is very prominent in some places, but it is not at all obvious in other places where the land to the east is high. However, the fault system, although not yet well defined everywhere, can be followed in a general way. Farther north the eastern scarp is recognizable as the mountain front on the west side of Lake Tahoe. Lake Tahoe, whose waters drain to the east, the Carson Range on its east side, and the forested and well-watered mountains both south and north of it are by common usage "in the Sierra Nevada." But to the geologist they are in fact in the Great Basin, and in the Basin and Range Province.

From its south end the crest of the Sierra Nevada as far north as Henness Pass (see figs. 37 and 38) is the Pacific drainage divide, or the watershed that separates the waters that flow into the Great Basin from those that flow into the Pacific Ocean. However, about a mile north of Henness Pass, which is near Webber Lake, the crest of the range and the Pacific drainage divide part company. The crest of the Sierra Nevada continues to the northwest, but the Pacific drainage divide turns east for 20 miles to skirt the headwaters of the Feather River. It turns north to Beckwourth Pass and continues northwesterly along the crest of the steep slope known as the Honey Lake scarp, past Susanville to a point about 10 miles north of Chester, where it again swings to the northeast. Along all of this course it divides the headwaters of the Feather River, which drains to the west, from the Great Basin.

From Henness Pass the crest of the range continues northward to Yuba Pass, thence to Haskell Peak and Mills Peak, whence it passes around the Lakes Basin to the south end of McRae Ridge. Along this course it divides the headwaters of the Yuba River from those of the Middle Fork of the Feather River. From McRae Ridge the crest may be thought of either as Bunker Hill Ridge through Pilot Peak or as Eureka Ridge. In either case the crest seems to end at the canyon of the Middle Fork of the Feather River below Nelson Point. But it does not really end there; rather, it is breached by the Middle Fork. The crest continues northwest of the Middle Fork along Bachs Creek Ridge to Clermont Hill, and it is the ridge that continues northwest of Clermont Hill, west of Meadow Valley, to Spanish Peak and Mt. Pleasant. Beyond Mt. Pleasant it is breached by the North Fork of the Feather River—Belden would be about under it—and 10 or 15 miles farther to the northwest the Sierra Nevada is terminated by burial beneath the younger volcanic rocks.

The eastern limit of the range north of Henness Pass is the base of the steep scarps along Sierra Valley, Mohawk Valley, Eureka Ridge, American Valley, and Meadow Valley. Beyond the North Fork of the Feather River the east base terminates at a still poorly defined point

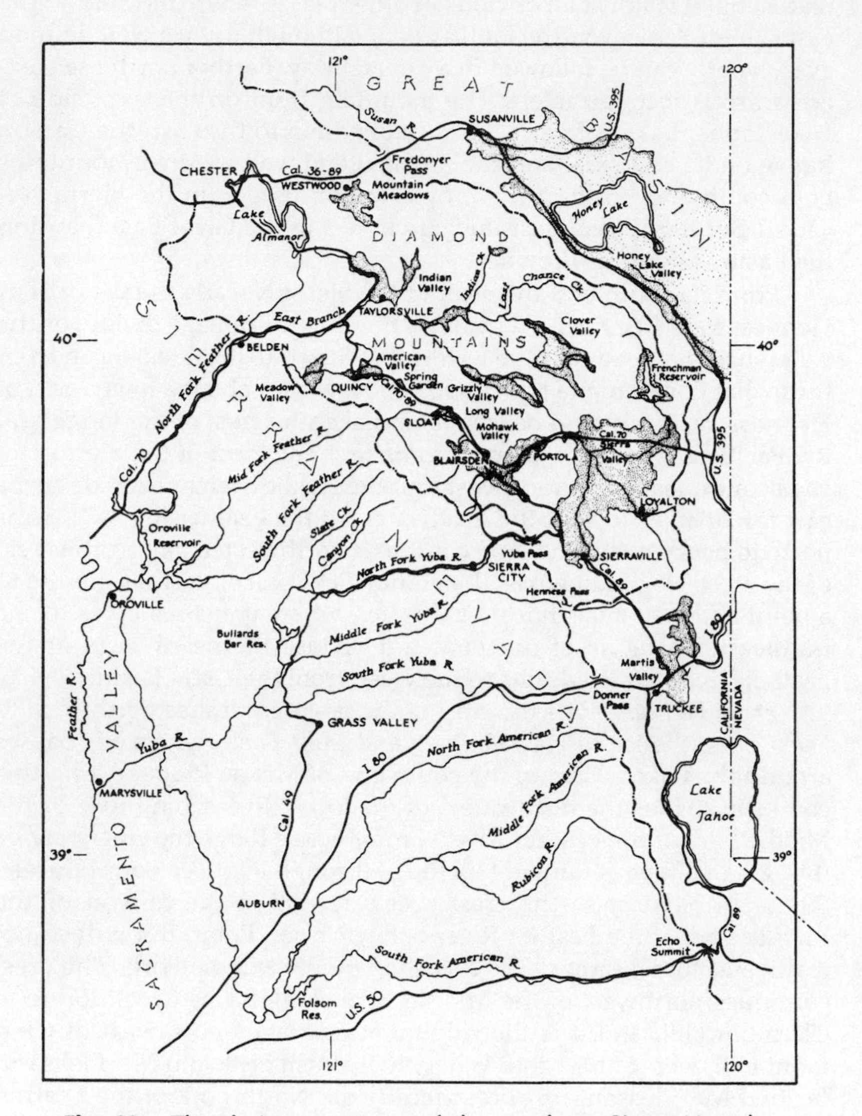

Fig. 38. The drainage pattern of the northern Sierra Nevada and Diamond Mountains and the extinct lakes. The main streams of the Sierra Nevada, except for the Rubicon River west of Lake Tahoe, flow directly to the west or southwest. Those of the Diamond Mountains flow to the northwest or southeast. The lake basins (patterned excepting Lake Tahoe) are east of the Sierra Nevada.

west of Lake Almanor. Thus all of the country beyond the east base of the Sierra Nevada that is tributary to the Feather River is in the Basin and Range Province, but not in the Great Basin. However, by custom and common usage it is considered to be a part of the Sierra Nevada.

The two main branches of the Feather River are the only Sierran rivers that breach the crest, and, whereas their courses west of the crest are directed to the southwest as are the other streams of the range, the courses of their headwater tributaries reflect the fault block pattern of the Basin and Range Province, as does the topography. The headwater tributaries have long reaches that flow either southeasterly or northwesterly (see figs. 37 and 38). The valleys and ridges are elongate to the northwest and are parallel to the main faults of the region. Many of the valleys contained lakes that have become extinct only in late geologic time (see fig. 38). Lakes occupied Sierra, Mohawk, and Grizzly valleys; Spring Garden; American and Indian valleys; the Almanor and Mountain Meadows basins, now occupied by reservoirs; Clover Valley; and Little Last Chance Valley, now occupied by Frenchman Reservoir.

Most of the valleys and some of the ridges in that part of the Basin and Range Province drained by the Feather River have names, but at present there is no name in common use for the entire region. It seems once to have been called the Diamond Mountains, a name that is useful and that deserves to be revived. In this text I apply it to all the country north of Sierra Valley as far as State Highway 36 and from the crest of the Sierra Nevada, as defined above, to the Honey Lake scarp.

Between the Sierra Nevada on one side and Grizzly Ridge and Keddie Ridge on the other (see fig. 126) there is a relatively low strip of land that extends from Sierra Valley near Calpine to American Valley and, perhaps, beyond Lake Almanor. Although smaller and less sharply bounded than the Owens Valley, it is analogous to that great depression far to the south. It needs a name and I have called it informally the Plumas Trench. It is a matter of more than passing interest, but not to this history, that a view to the northwest along the trench reveals Lassen Peak as the centerpiece.

The western slope of the Sierra Nevada in the central and northern parts rises very gradually from the Great Valley with but a few interruptions to a maximum at the crest, which is close to the east base. In the north, the high peaks along the crest are a little more than 8,000 feet high. The general height of the crest increases to the south, culminating at Mt. Whitney, whose altitude is 14,496 feet. South of Mt. Whitney the crest is progressively lower until at Walker Pass its altitude is only 5,250 feet.

The main rivers excepting only the Kern flow rather directly down the western slope into the Great Valley. The Kern, however, flows due south along a fault for 38 miles before it turns west to enter the Valley near Bakersfield. All the large streams are in sharply V-shaped canyons whose maximum depth of several thousand feet is reached near the middle of their courses. Between the canyons are plateau-like areas with the quite gentle relief of an old erosion surface, in sharp contrast to the steep walls of the canyons (see fig. 130).

The streams of the east side of the range are short, small, and have steep gradients. The largest are the East and West Walker rivers, the Carson River, and the Truckee River, which flows out of Lake Tahoe. All the drainage basins on the east are small compared to those of the rivers in the west slope. The character of the streams on the east slope reflects the recent uplift of the range.

The Plumas-Eureka State Park and the recreational areas of the national forest that immediately surround it are on or east of the crest of the Sierra Nevada as it is defined geologically. Mohawk Valley and the mountainous lands across the valley, Penman Peak, and all of Grizzly Ridge and the lands beyond are in the Basin and Range Province, but not in the Great Basin. However, there can be little doubt that the common usage that places all of that region in the Sierra Nevada will continue to prevail.

This great block of the earth's crust that we call the Sierra Nevada and the Diamond Mountains is made up of three distinctly different groups of rocks. The oldest of these, described in Chapter 3, were once mostly layered sedimentary and volcanic rocks deposited in the sea. They are the rocks of the Nevadan geosyncline, which is the western part of the greater Cordilleran geosyncline. Their ages span the time from late Ordovician or some time early in the Silurian period to almost the end of the Jurassic period, some 250 million years. All of them are deformed and metamorphosed—they are now metamorphic rock. The layering, which in many places is quite conspicuous, stands nearly vertical or is steeply inclined. Most of the rock is rather dark colored, mostly gray or green, or stained brown by iron oxide as a consequence of weathering. The rocks are well displayed in the deep canyons and are easily accessible along State Highway 70 between Oroville and Quincy.

One of the best places to become acquainted with them is in and near the Plumas-Eureka State Park, and in the Lakes Basin and Sierra Buttes recreational areas, where they have been divided into easily recognizable formations—groups of rocks with similar characteristic features. There the rock layers and the formations that they comprise are inclined to the east at an average of about 60 degrees from the horizontal, like a giant layer cake tilted up and sliced through. If one wished to, one could walk across the edges of the rock layers with few interruptions from Onion Valley on the west to the Mohawk Saddle between Mills Peak and Haskell Peak on the east.

Some metamorphosed igneous intrusions associated with the layered rocks are described in Chapter 3 along with them, as is the rock serpentinite, which is thought to be derived from the earth's mantle and emplaced in connection with the process of subduction.

The second great group of rocks are the granitic rocks that comprise the plutons, described in Chapter 5. These, too, are well exposed along State Highway 70 between Pulga and Belden, where for 16 miles the North Fork of the Feather River and the highway are within the boundaries of the Grizzly and Bucks Lake plutons. Granitic rocks are also present in eastern Plumas and Sierra counties, but there they are so much weathered that many of their features are obscured.

The granitic rocks are massive—they are not layered. They are granular, speckled black and white or stained brown as a result of weathering.

The rocks of the first group suffered an almost paroxysmal deformation and metamorphism near the end of the Jurassic period in an important geologic event known as the Nevadan Orogeny, which is the subject of Chapter 4. The layered rocks were complexly folded and faulted. The invasion of magma to form the granitic rocks of the second group began at that time and continued into the Cretaceous period. This complex event of deformation, metamorphism, and invasion of magma resulted in the creation of a folded mountain system known as the Ancestral Sierra Nevada.

The Ancestral Sierra Nevada was a terrain vastly more extensive than the present Sierra Nevada. It was about 200 miles wide, extending from central Nevada as far west as the center of the Great Valley. To the north it extended into Oregon and possibly as far as Alaska. On the south it extended into Mexico. The rocks of the first and second groups together have been called the Subjacent Series, or, the basement rocks, or simply the basement. They are called this because they form a basement or foundation upon which lie the rocks of the third group, often called the Superjacent Series.

The rocks of the third group, the Superjacent Series, are younger than those of the other two. They rest on a surface of unconformity eroded during a long time interval after the Nevadan Orogeny. Consequently they rest on the upturned edges of the metamorphic rocks, and on the granitic plutons alike.

Early in the history of this group, the sea bordered the Sierra Nevada along its present western edge, and some sedimentary rocks of marine origin are thus included in the Superjacent Series. But the marine sedimentary rocks belong essentially to the province of the Great

Valley. Their counterparts are unknown in the Sierra Nevada or in the region immediately east of it.

Most of the rocks of the Superjacent Series are layered sedimentary and volcanic rocks deposited on land above sea level, but some igneous intrusions that fill source vents for the volcanic rocks are included with them.

The principal rock types are gravel, sand, and clay and their lithified equivalents conglomerate, sandstone, and shale; lava flows; volcanic mudflow breccia accompanied by volcanic conglomerate, sand, and mud; volcanic ash, mostly welded ash flows; and, among the youngest rocks, the still soft sediments deposited in the recently extinct lakes, and the till of glacial moraines. These are described in Chapters 6, 7, and 8.

Unlike the layered rocks of the Subjacent Series, those of the Superjacent Series are not metamorphosed and their layers are still in their original nearly horizontal position. They have been deformed only slightly and locally, as, for example, adjacent to faults.

Places where the significant features of the rocks can be seen are described in the text.

PART II

The Events Leading to the Ancestral Sierra Nevada

3

The Sedimentary and Volcanic History of the Nevadan Geosyncline

INTRODUCTION

The story of the Ancestral Sierra Nevada begins about 400 million years ago when the western margin of North America was in western Utah, as shown in fig. 39, nearly 600 miles east of where it is today. The margin or edge of the continent is not the shoreline but is that rather narrow region where the lighter-weight (lower-density) rocks of the continent meet the heavier (higher-density) rocks of the ocean basin. The ocean shore was still farther east because the oceans lap onto the continents.

A span of ocean floor perhaps 700 miles wide at that time received in its eastern half large amounts of the nonvolcanic sediments limestone and dolomite, which are of organic origin, and lesser amounts of sand and mud derived from the continent to the east. The western part, which became the Nevadan geosyncline, also shown in fig. 39, received enormous amounts of volcanic rock erupted through the sea floor, with only much smaller amounts of the organic sediments chert and limestone, and nonvolcanic sand and mud derived from a part of the continent that lay to the southeast.

Continued volcanic activity in the western part of this section of the ocean, an area that is now 200 miles wide and 3,000 miles or more long and which extended from Alaska into Mexico, undoubtedly constructed large submarine volcanoes, some of which may have become islands, although proof of islands in the northern Sierra Nevada is lacking. Modern analogues of this situation are the islands of Indonesia, the Aleutian Island chain, and other arcuate chains of volcanic

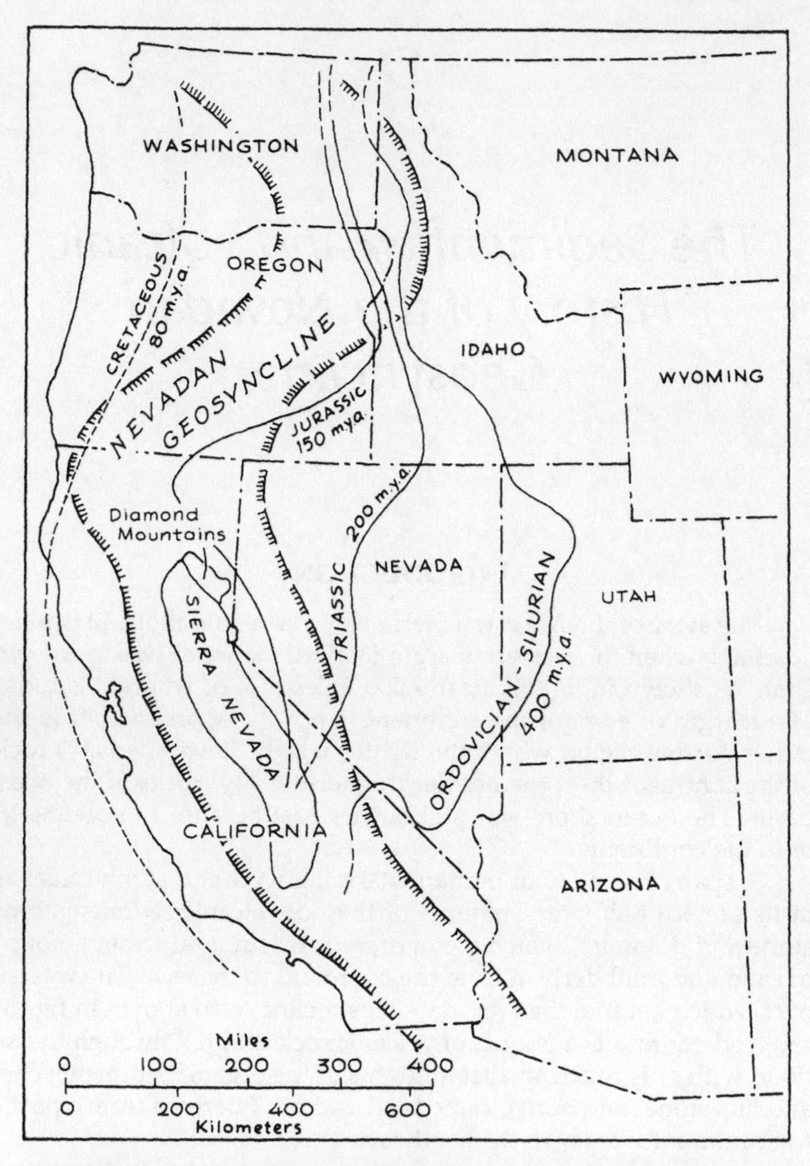

Fig. 39. The Sierra Nevada and Diamond Mountains in relation to the western edge of the continent. The position of the latter is shown as it was at about 400, 200, 150, and 80 million years ago. The hachured lines mark approximately the present boundaries of the Nevadan geosyncline.

islands that festoon the western Pacific Ocean basin. They are called island arcs although islands are not a necessary part of the structure. In terms of the concept of plate tectonics, subduction takes place on the oceanic side of the arcs, melting occurs above the subduction zone, and magma rises through the rock above the subduction zone to produce volcanic activity, as shown in fig. 3.

The accumulation of sedimentary and volcanic rock along the western side of North America, interrupted from time to time by episodes of deformation, continued for more than 250 million years, during which the rocks of the Nevadan geosyncline accumulated. All the ancient sedimentary and volcanic rocks deposited in the sea were layered, or bedded, and the layers were about horizontal when deposited. Later, about 150 million years ago, the rocks were deformed, that is, they were folded, faulted, and metamorphosed during the Nevadan Orogeny, which terminated the long period of deposition. The deformation resulted in reduction of the width of the geosyncline by about one-half. In the northern Sierra Nevada and Diamond Mountains the layers are now steeply inclined and occur as long strips of rock. The rocks are not all alike, which allows them to be divided into formations.

Presently the geosynclinal rocks of the region fall naturally into four parts: an eastern and a central belt of metamorphic rocks separated by the long, narrow strip called the Feather River Serpentinite, and a western belt of somewhat different character called the Smartville Complex, as shown in fig. 40.

The rocks of the eastern belt have been studied more thoroughly than those farther west and are therefore better known. They are easily accessible along trails, roads, and stream beds, especially in the region between the North Fork of the Yuba River and the Middle Fork of the Feather River in eastern Plumas and Sierra counties, where, because they were laid bare by glaciation, they can be examined in detail. This is the region of the established recreation areas around Sierra Buttes, Gold Lake, the Lakes Basin, and the Plumas-Eureka State Park.

The rocks of the central and western belts are not so well exposed to view for the reason that, not having been glaciated, they are beneath a cover of forest and brush and much of the rock is deeply weathered (decayed). Although fresh, unaltered rock is present in the deep canyons, large segments of the canyons are now underwater behind the large dams. Other portions of the canyons are physically difficult to reach. Only the North Fork of the Yuba River and the North Fork of the Feather River are easily accessible. In spite of the difficulties, the rocks of the western and central belts are now the object of intense study by many geologists and good progress is being

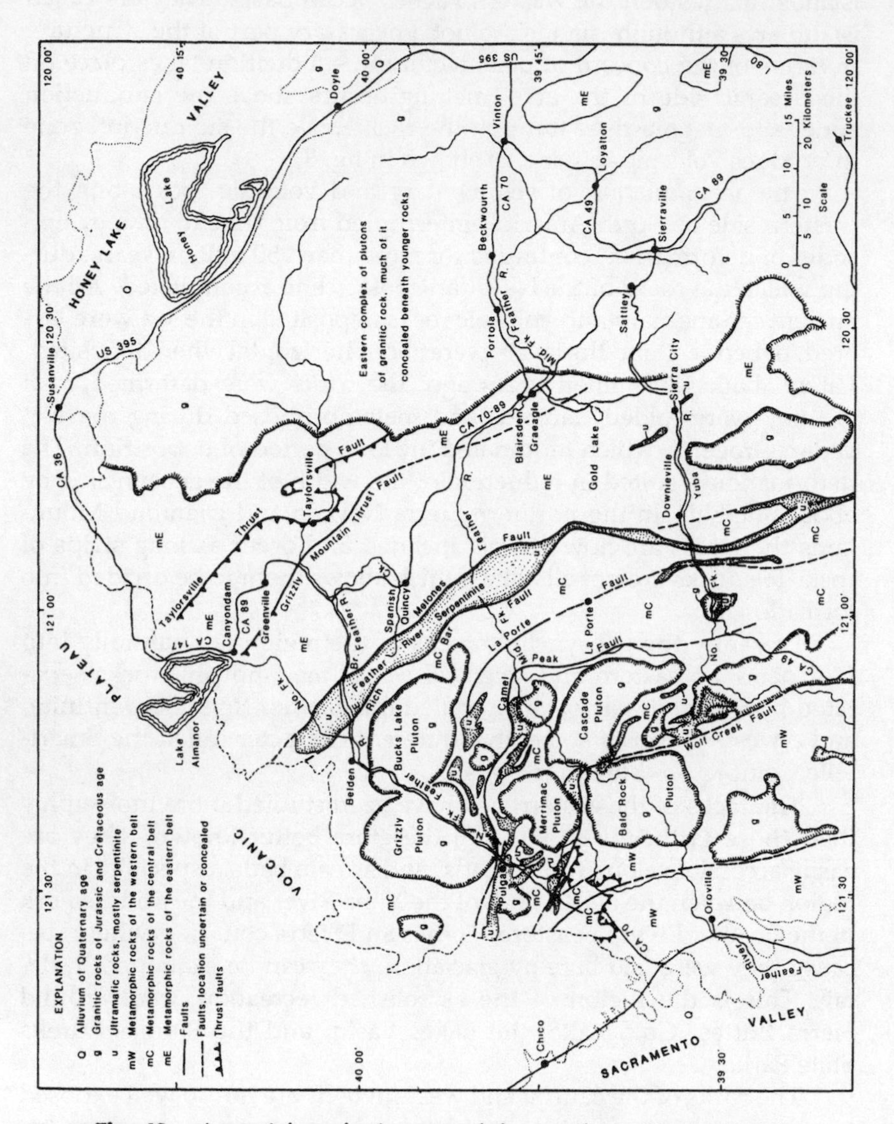

Fig. 40. A partial geologic map of the northern Sierra Nevada and Diamond Mountains. The rocks of the Subjacent Series are divided into the eastern, central, and western belts of metamorphic rocks, the Feather River Serpentinite, and the eastern and western groups of plutons of granitic rocks. The rocks of the Superjacent Series are omitted.

made in understanding the region. For these reasons I have described first the rocks of the eastern belt, which should aid the reader in understanding the less well understood rocks farther west.

The formations of the eastern belt in Plumas and Sierra counties, especially in and near the recreational areas, are inclined to the east at an average slope of about 60 degrees, and, excepting areas where folding is intense, they continue in about that same position from the vicinity of Lake Almanor on the north to the North Fork of the American River south of Interstate Highway 80. As shown in fig. 41, they are progressively younger from west to east although they are repeated (duplicated) on the two sides of the Grizzly Mountain Fault. Thus the rocks of the eastern belt are divisible into two parts, almost but not quite identical in character. A principal difference is that the Arlington Formation of the western part appears to be the equivalent of the Goodhue, Reeve, Hosselkus, and Swearinger formations of the eastern part. Rocks of Jurassic age are not present in the western part, but are immensely thick in the eastern part.

The metamorphism that affected the rocks of the northern Sierra Nevada is of such low intensity that the features that permit the original nature of these rocks to be determined are well preserved. For that reason, in the descriptions that follow, I have, in most instances, omitted the prefix "meta" from the names appropriate to them before they were metamorphosed. The only metamorphic rock names used are slate and schist. However, reminders of the fact of metamorphism are inserted here and there so that there should be no confusion about which rocks are metamorphosed and which are not.

ROCKS OF THE EASTERN BELT

Shoo Fly Formation

The rocks of this, the oldest formation of the northern Sierra Nevada, were deposited during the Silurian or possibly even during the late part of the Ordovician period; the age is based on fossils found in limestone on the northeast slope of Mt. Hough about 3 miles south of the town of Taylorsville in Indian Valley, and on some microscopic organisms called radiolaria that were found near the Spencer Lakes.

The rocks of the Shoo Fly Formation are those so well exposed in road cuts and stream banks along State Highway 70 between Virgilia and the Y that is the junction of State Highways 70 and 89 near Paxton; along State Highway 89 northerly from the Y to Indian Falls; and to the southeast from the Y to Quincy. In this region the most conspicuous rock is slate derived from mud, but sandstone is also abundant. Layers several feet thick grade from sand at the bottom to what was

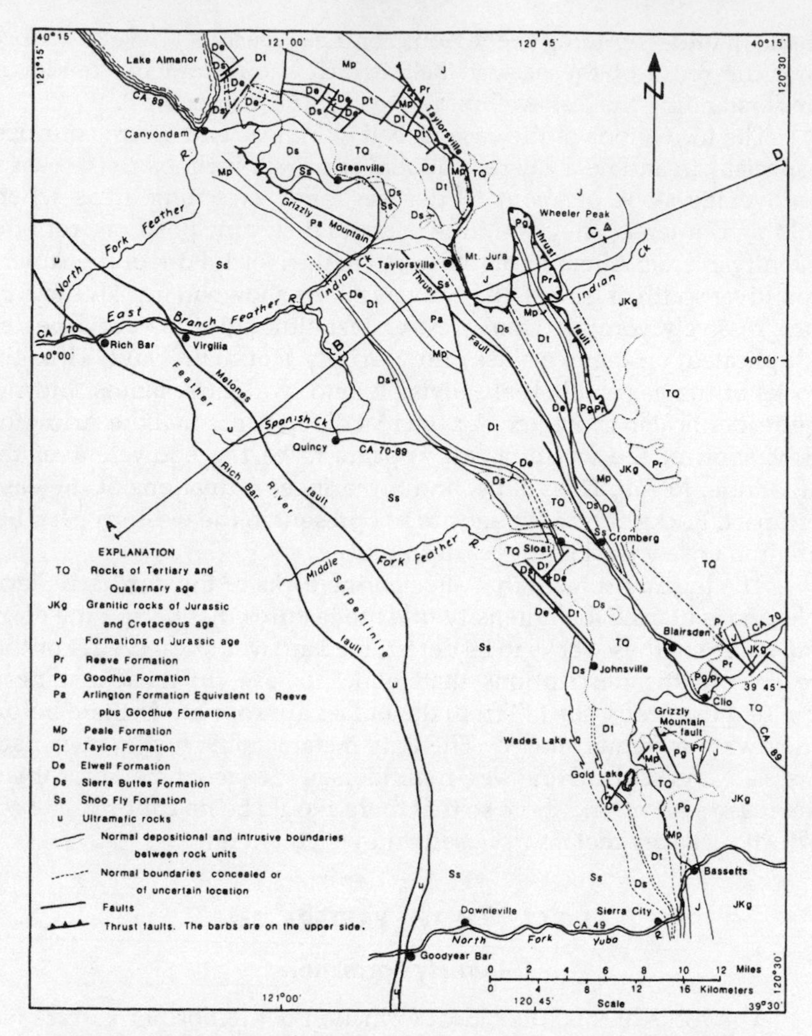

Fig. 41. Geologic map of the metamorphic rocks of the eastern belt showing also the position of the cross sections of figs. 73 and 74. Compiled from the published and unpublished studies of J. A. D'Allura, C. Durrell, J. L. Hannah, L. Robinson, and V. E. McMath.

mud, and is now slate, at the top. An excellent place to see this relationship is in the bed of Indian Creek near the Y. Small bodies of limestone are present also, but none show in the road cuts. Chert is also an important rock in the Shoo Fly Formation.

The same rocks of the Shoo Fly Formation are present along the Quincy–La Porte Road between Nelson Point and Onion Valley, and along State Highway 49 between Sierra City and Goodyears Bar. Most of the rock of the Shoo Fly Formation are sand and mud derived from a continental land mass, but volcanic rocks do occur along its eastern edge.

One of the best places to examine the Shoo Fly Formation is in and close to the Plumas-Eureka State Park along Big Jamison Creek and the Johnsville–La Porte Road. Traveling along that road from Johnsville or from the park campground, one first encounters the Shoo Fly where the road enters the forest after passing along the shelflike way overlooking the meadow. Here the rock is a rust-stained black slate derived from mud. Next to that is gray dolomite, a granular rock akin to limestone, but containing much magnesium as well as calcium. Although the dolomite was probably of organic origin, no recognizable remains have been found in it.

Beyond this point it is difficult to describe where different kinds of rocks appear, but westward from the dolomite along the road there is first a dark green mafic volcanic ash, followed by light-colored silicic volcanic ash that contains large crystals (phenocrysts) of quartz embedded in a fine-grained matrix. At the state park boundary there is excellent slate (see fig. 29) with very perfect cleavage. Here also one can look up at the south slope of Eureka Peak and see patches of light-colored rock seemingly spaced at random over the slope. These are masses of dolomite like that along the road that have flowed plastically into the axes of folds during metamorphism (fig. 42).

For some distance beyond the state park boundary, the rocks are mostly metamorphosed silicic volcanic ash with distinctive glassy crystals of quartz embedded in a fine-grained matrix. In some places, the rocks have cleavage, but in other places they do not.

Along the road in the southwest corner of Sec. 27 there is a layer of serpentinite only a few hundred feet wide. The rock appears blackish, but it is dark green on freshly broken surfaces. Surfaces long exposed to weathering are brown. A layer of light-colored talc about a foot thick is present along the west margin of the serpentinite.

Beyond the next right-hand turn of the road, where the grade is steep and the cliff on the south side of the creek seems almost to overhang, there is a mass of grayish green chert whose distinct layers one to two inches thick are folded on a small scale (fig. 43). The rock contains the poorly preserved remains of the marine organisms radiolaria, but they can be seen only with a microscope. Westerly from the chert most of the rock is silicic volcanic ash with distinct quartz crystals. At Rene's Falls quartz crystals (phenocrysts) of exceptional size, some as much as three-fourths of an inch across, which show the typical square and hexagonal cross sections, are abundant in the light-colored volcanic ash. One can see there that in some layers the large

Fig. 42. The south slope of Eureka Peak as seen from the Johnsville–La Porte Road at the Plumas-Eureka State Park boundary. Light-colored masses of dolomite are surrounded by darker slate, chert, and volcanic rocks of the Shoo Fly Formation. The dolomite was originally in sheetlike layers (beds), but during the metamorphism it flowed plastically to accumulate in the axes of folds. The folds have amplitudes of tens to a few hundreds of feet. They are similar to those shown in figs. 49 and 50.

quartz crystals are graded in size across a bed. From that fact one can conclude with confidence that the rocks were deposited in water.

Above Rene's Falls there is more dark slate in which can be seen tightly appressed folds like those of a closed bellows (fig. 44*a*; see also fig. 11). Layers of light gray sandstone composed almost entirely of quartz are present in the slate. Such rock is called quartzite.

This is the limit of the volcanic part of the Shoo Fly, and the end of well-exposed rock, but the Shoo Fly Formation is present as far west as the serpentinite at Whiskey Creek. Slate and quartzite are the dominant rocks west of Rene's Falls, but chert is present at many places. Two layers of gray limestones are exposed in road cuts, one on each side of McRae Ridge, but neither has yielded fossils. Nor have any radiolaria well enough preserved to indicate the age of the formation been found in the chert close to the road. Very coarse sandstone with small pebbles is present along the road in Sec. 21, T. 22 N., R. 10 E., a mile east of Whiskey Diggings.

Fig. 43. Radiolarian chert of the Shoo Fly Formation. Folded layers 1 to 2 inches thick of grayish green chert. On the Johnsville–La Porte Road along Big Jamison Creek. Near the S 1/4 Cor., Sec. 27, T. 22 N., R. 11 E.

That the rocks of the Shoo Fly Formation are surely marine is indicated by the radiolaria in the chert, which also indicate that the water was perhaps several thousands of feet deep. The association of the silicic volcanic ash with radiolarian chert means that the ash is also a deep-water deposit. Possibly the eruptions were submarine, but no possible vents have been recognized. The sand and mud pose no special problem of origin, because they could easily be transported by currents into deep water. They do require a land mass as a source, and, while its location has not yet been identified for certain, it probably lay far to the southeast.

Because the rocks of the Shoo Fly Formation have been severely folded as is demonstrated by figs. 11, 43, 44, and 50, it is not yet clear whether the layers on the whole face eastward or westward. The width of the Shoo Fly Formation, east to west, is about 7 miles. Nothing can be said for certain about its thickness, but it must surely have exceeded 10,000 feet and possibly was much thicker than that. To the north the Shoo Fly extends into the region west of Lake Almanor and to the south it extends far beyond the North Fork of the American River.

a

Fig. 44. **a,** Nearly isoclinal folds in slate of the Shoo Fly Formation. In the bed of Big Jamison Creek above Rene's Falls. The rock surface is underwater. SW 1/4, Sec. 28, T. 22 N., R. 11 E. **b,** Diagrammatic explanation of **a.**

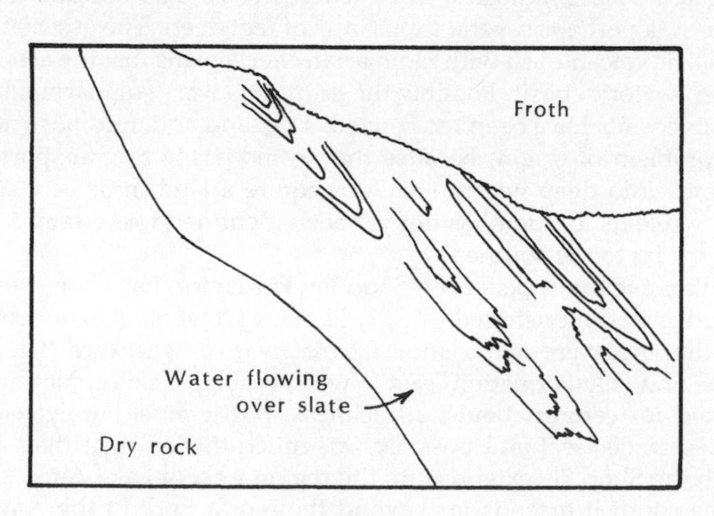

b

Fig. 45. Volcanic breccia of the Sierra Buttes Formation. Note pencil for scale. On the Johnsville–La Porte Road 0.25 mile southwest of the entrance to the Plumas-Eureka State Park Campground. SW 1/4, Sec. 26, T. 22 N., R. 11 E.

Sierra Buttes Formation

The Sierra Buttes Formation lies east of the Shoo Fly Formation and is the next youngest. It forms the spectacular peaks of Sierra Buttes, from which it was named, the summits of Mt. Washington and Eureka Peak, and the nearly white east-facing cliffs between Mt. Elwell and Sierra Buttes.

Between Eureka Peak and Sierra Buttes, the formation is almost exclusively silicic volcanic ash and breccia. The rock is light green when freshly broken but exposed surfaces are mostly nearly white, pale bluish gray, or tan. In some places it is rusty with iron oxide produced by weathering, and this is especially the case where it is under forest. The fragmental character of the original fine-grained ash is not recognizable without the help of a microscope, but the large fragments of breccia, one inch and larger, are easily seen and are present nearly everywhere. Good examples occur along the Johnsville–La Porte Road about one-fourth mile south of the entrance to the Plumas-Eureka State Park campground, as shown in fig. 45. Breccia is easily recognized high on Eureka Peak, on Mt. Washington, and especially by the trail from the Lakes Basin campground to Mt. Elwell, in the SE 1/4, Sec. 12, T. 21 N., R. 11 E. (figs. 46 and 47). Many of the blocks of the breccia show flow layering. The rather large crystals (phenocrysts) of quartz (one-eighth to one-quarter inch in diameter) are visible nearly everywhere, and in many places dull white rectangular crystals (phenocrysts) of feldspar are also present. Large fragments of pumice are recognizable farther north near the summit of Grizzly Ridge.

Cleavage is present nearly everywhere but is not conspicuous in the Lakes Basin and park areas, where much of the rock is only slightly weathered at the surface. Where the rock is more weathered the cleavage is prominent and the rock splits into little lenses, each likely to contain one to several quartz phenocrysts.

Black radiolarian chert is sparingly present and can be readily visited on the west side of Rock Lake and on a prominent knob on the slope southeast of Jamison Lake. The presence of radiolarian chert close to both the top and the bottom of the formation indicates that it was entirely marine and that the water was quite deep.

The lowest bed (the basal bed) of the Sierra Buttes Formation is a conglomerate that contains rock fragments recognizable as derived from the Shoo Fly Formation, upon which it rests. It can be easily reached and studied on the ridge just southwest of the south peak of Eureka Peak (fig. 48). A better and more informative display is present on the slope south of Wades Lake, where the basal bed can be followed around an anticline, an archlike fold, and through the adjacent syncline, a troughlike fold, as shown in figs. 49 and 50. On the anticline, the Sierra Buttes Formation rests on dolomite of the Shoo Fly, and its basal bed contains abundant fragments of the dolomite. In the syncline and continuing upward, the basal bed rests on slate and silicic volcanic rocks like those along Big Jamison Creek and it contains fragments of those rocks. Thus the basal bed contains and rests on different kinds of Shoo Fly rocks. The surface on which the basal bed rests is called an unconformity and it signifies several important things.

If in the mind one unfolds the Sierra Buttes Formation to its original horizontal position (it can be done on paper also), one discovers that the beddings of the Sierra Buttes and Shoo Fly formations are not parallel, but meet at an angle of about 35 degrees. This relationship proves that the Shoo Fly Formation was deformed, either folded or tilted, before the Sierra Buttes Formation was deposited on it. It shows also that the Shoo Fly Formation was brought near to or above sea level and eroded to produce the surface of unconformity, on which different kinds of rocks were exposed. Only in this way could the various kinds of rocks that are incorporated into the basal bed of the Sierra Buttes become available. Because the Sierra Buttes Formation is ma-

Fig. 46. Cliffs of nearly white silicic volcanic breccia of the Sierra Buttes Formation. The cliff is the headwall of a cirque in the SE 1/4, Sec. 12, T. 21 N., R. 11 E., 1.5 miles southwest of the Lakes Basin Campground by way of the trail to Mt. Elwell. It is also a quarter mile west of the junction of trails at the top of the slope west of Silver Lake.

Fig. 47. Coarse volcanic breccia of the Sierra Buttes Formation showing crude bedding. On the floor of the cirque shown in fig. 46. Phenocrysts of clear glassy quartz and dull white feldspar are abundant. Some of the blocks are flow banded.

Fig. 48. The conglomerate at the bottom of the Sierra Buttes Formation. Beside the old wagon road west of the south peak of Eureka Peak, Plumas-Eureka State Park.

rine in origin, the Shoo Fly Formation must then have been brought below sea level for the Sierra Buttes to be deposited on it in quite deep water.

This remarkable display shows also, of course, that the Sierra Buttes and Shoo Fly formations were folded together into the anticline and syncline at a later time—that is, during the Nevadan Orogeny, the subject of Chapter 4. That orogeny affected all the metamorphic rocks younger than the Sierra Buttes Formation, which are described farther on. Of course, this means also that the rocks of the Shoo Fly Formation were deformed twice.

As a consequence of the second deformation, the Sierra Buttes Formation is inclined to the east at about 60 degrees from the horizontal. It does not reappear again to the east, and no doubt it terminates at depth against the plutons of granitic rock that are present there. To the west its continuation would be above ground where it has been eroded away. To the south it extends as far as the North Fork of the American River but it is thinner there. It is thin in the vicinity of Taylorsville and Indian Falls, but it is thicker again farther north on Keddie Ridge. Its total known length is about 75 miles. In the Lakes Basin the Sierra Buttes Formation is about 4,000 feet thick. Its average thickness is not known, but if it is as much as 2,000 feet, and taking into

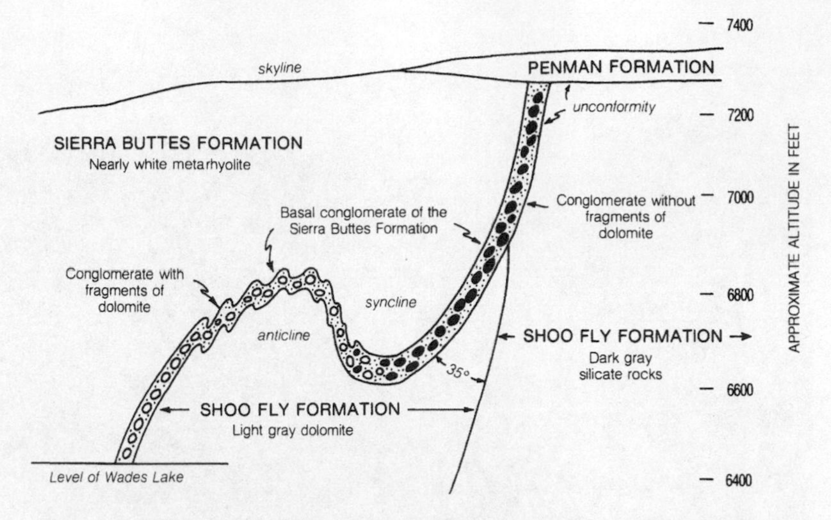

Fig. 49. A sketch showing the folds and the unconformity between the Shoo Fly and Sierra Buttes formations at Wades Lake. The fold is evident in the configuration of the bottom bed (the basal conglomerate) of the Sierra Buttes Formation. The amplitude of the fold—the distance measured vertically between the arch (the anticline) and the trough (the syncline)—is about 200 feet. Smaller folds on the crest of the anticline are conspicuous. The angle between the layers of the two formations is about 35 degrees. The surface of contact between them is the result of erosion and is called an unconformity.

account that it might easily extend as far as 5 miles below the surface and that perhaps as much as half of it has been lost to erosion, it is possible that its original volume was as much as 300 cubic miles. This is, indeed, a very large amount of silicic volcanic rock.

No vents that could be the source of the Sierra Buttes volcanic rocks are known to me. One place on Sierra Buttes and another on Grizzly Ridge near Argentine Rock have been suggested as having been source vents but I find neither of them to be convincing. Vents may not be visible at the present erosion surface, the position of which is merely accidental as far as the original distribution of the Sierra Buttes Formation is concerned. All vents could have been above the present erosion level and have been eroded away, or they could be below the present erosion surface, where they remain concealed.

The Sierra Buttes Formation is, of course, younger than the Shoo Fly Formation because it is above it. Radiolaria from chert near the bottom of the formation west of Greenville indicate a late Devonian age, similar to that of the overlying Elwell Formation.

a

Fig. 50. **a,** The anticline and syncline at Wades Lake as seen from the south slope of Mt. Washington. **b,** Diagrammatic explanation of **a.**

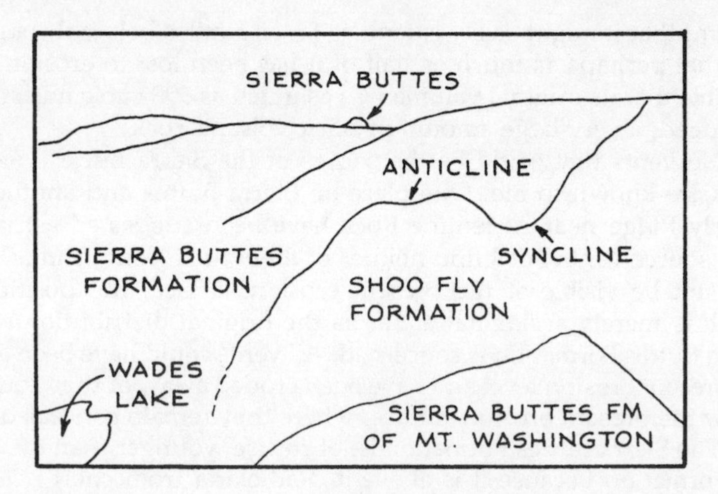

b

Fig. 51. Elwell Formation. Alternating beds of volcanic ash and black radiolarian chert. Each layer of ash is a graded bed. Both ends of the pencil are on ash. Exposed in the south wall of the cirque shown in fig. 46. See fig. 46 for location.

Elwell Formation

The Elwell Formation, named for Mt. Elwell, rests on the Sierra Buttes Formation and, like it, is inclined to the east. In the Lakes Basin District, where it can be easily examined, it consists mostly of dark gray to black radiolarian chert in layers 1 to 2 inches thick. Streaks and nodules of grayish white phosphate rock enclosed in the chert contain well-preserved radiolaria. Both silicic and andesitic volcanic ash are interlayered with chert (fig. 51). Graded beds of the ash show that the tops face to the east. A lava flow of andesite is present in the formation southwest of Long Lake in Sec. 12. Its fragmented top is cemented with nearly black chert. Sills of andesite with columnar structure intrude the Elwell in the same neighborhood. Small folds are present nearly everywhere in the Elwell Formation, a feature rather typical of radiolarian chert wherever it occurs.

These aspects of the Elwell Formation are well displayed on the two ridges south of the southwest corner of Long Lake, west of the trail to Mt. Elwell, on the divide southwest of Mt. Elwell, and along the Mt. Elwell trail between the divide and Mud Lake.

The Elwell Formation is thick in some places and thin in others.

Southwest of Long Lake, it is about 300 feet thick, about half of which is chert. However, the chert represents slow accumulation of the microscopic radiolaria on a deep ocean floor so that the span of time represented by the Elwell may be many times that which would be required to account for an equal thickness of fragmental volcanic rock. On Keddie Ridge, the Elwell is 6,000 feet thick, mostly of volcanic material.

The age of the Elwell is known to be late Devonian from fossils found near Dugan Pond about a half mile east of Packer Lake in Sec. 8, T. 20 N., R. 12 E. The fossils are ammonoids, ancient relatives of the living pearly nautilus. The age is confirmed by radiolaria from the nodules of phosphate rock in black chert south of Long Lake.

Because it contains both silicic and andesitic volcanic debris the Elwell Formation should be thought of as transitional from the Sierra Buttes Formation below it to the younger Taylor Formation, which overlies it.

Taylor Formation

The Taylor Formation, next younger than the Elwell, is almost all andesite. It contains some lavas with pillow structure, and volcanic sandstone is abundant, especially in the lowest part (figs. 52 and 53), but mostly it is volcanic breccia.

The original layering (bedding) can be seen at many places, notably in the volcanic sandstone. Its inclination to the east ranges between 15 degrees and vertical and averages about 60 degrees. Graded beds, cross-bedding, and small filled channels present at many places show that the tops of the beds face to the east.

The lowest part, which is easily reached on the ridges west of the south side of Long Lake, is mostly massive andesitic sandstone (see fig. 52) with only a few layers of fine-grained, thinly bedded rocks. A zone of well-bedded rocks is exposed along the west side of Long Lake and at the southwest corner of Silver Lake. Cross-bedding, scour channels, and graded beds shown in figs. 54–56 are well displayed at the latter place, which is easily reached by a trail that skirts the east and south sides of the lake.

Above or east of this strip of well-bedded rocks is a great thickness of breccia, some of which contains blocks as large as 2 and 3 feet across.

Some of the coarse breccia has abundant fine matrix, as shown in fig. 57, which portrays a smoothed and grooved glaciated surface at the junction of the Mt. Elwell and Long Lake trails, about a third of a mile west of the end of the road in the Lakes Basin campground. Some coarse breccia has a calcareous matrix that on weathering leaves

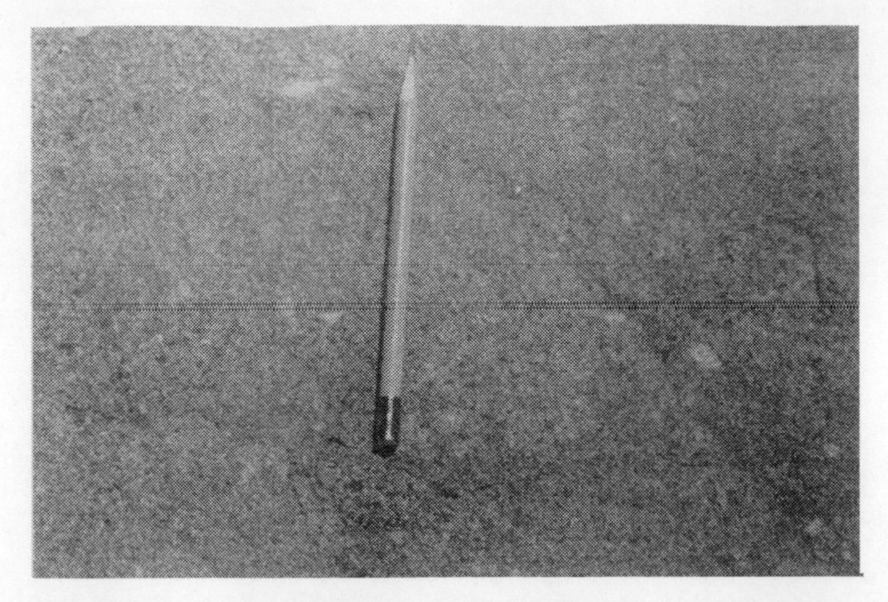

Fig. 52. Taylor Formation. Volcanic sandstone composed of broken crystals and pieces of fine-grained andesite with a few scattered small blocks of andesite. This is the lowest part of the formation, near Helgramite Lake, west of Silver Lake, Lakes Basin District. SE 1/4, Sec. 12, T. 21 N., R. 11 E.

Fig. 53. Taylor Formation. Volcanic sandstone (andesitic) with blocks larger and more abundant than in the rock of fig. 52. The same locality as fig. 52.

Fig. 54. Taylor Formation. Graded beds of andesitic sandstone and siltstone. The top is to the right. At the end of the trail along the south side of Silver Lake, Lakes Basin District. NW 1/4, Sec. 18, T. 21 N., R. 12 E.

Fig. 55. Taylor Formation. Cross-bedding in volcanic sandstone. The top is to the left. The same location as in fig. 54.

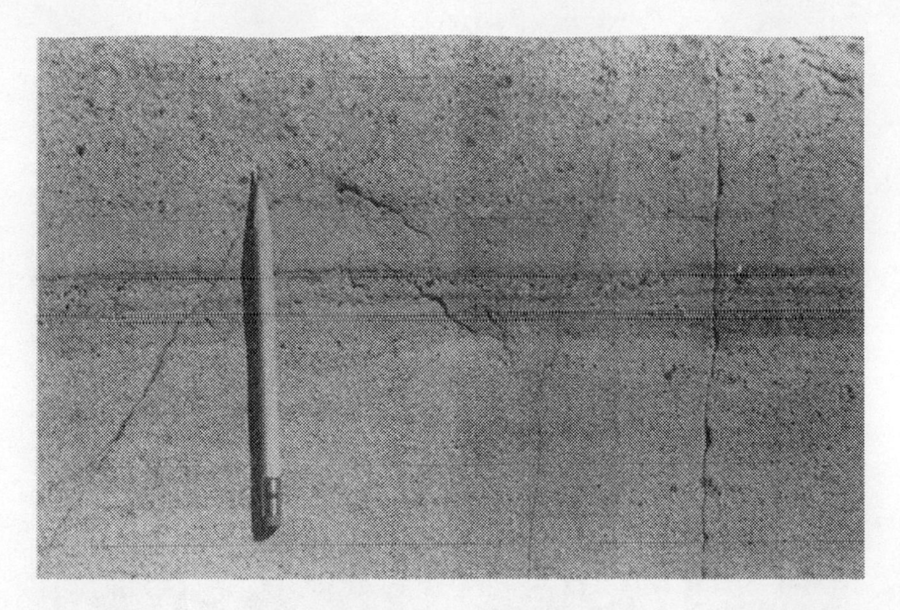

Fig. 56. Taylor Formation. A cut-and-fill channel in volcanic sandstone is immediately to the right of the pencil. The top is upward. The same location as in fig. 54.

Fig. 57. Taylor Formation. Volcanic breccia with blocks as large as 8 inches. The blocks are only slightly deformed but the matrix has cleavage (see also fig. 31). A conspicuous outcrop of rock with glacial striations at the junction of the Mt. Elwell and Long Lake trails, a half mile west of the Lakes Basin Campground. At the center of Sec. 7, T. 21 N., R. 12 E. The scale is given by the pencil.

Fig. 58. Taylor Formation. Coarse volcanic breccia. The matrix of the breccia contains the mineral calcite, which is calcium carbonate. The calcite dissolves readily, thereby leaving the blocks standing in relief. The blocks are sharply angular and unsorted. This huge piece of rock is not in place but was transported and deposited here by the glacier that filled the valley. About 0.75 mile from the Lakes Basin Campground by the trail to Mt. Elwell and Silver Lake. The handle of the hammer is 13 inches long.

an open or porous matrix. This makes the nature of the breccia strikingly clear, as shown in fig. 58.

Most of the blocks of the breccia of the Taylor Formation are porphyritic with phenocrysts as large as a half inch, of glassy green augite, or of duller dark green hornblende, which is the metamorphic alteration product of the original augite. Visible phenocrysts of white feldspar are uncommon. Ovoid to spherical amygdules—steam bubbles that later became filled with minerals, mostly milk white quartz—are visible nearly everywhere. Amygdules filled with dark green minerals such as chlorite and epidote are less conspicuous but also abundant.

With the aid of a microscope the finer matrix between the large blocks is seen to be composed in part of scoriaceous fragments, some of which are also amygdaloidal. Some fragments were vesicular glass in which the bubbles collapsed. Because the matrix was mechanically weaker and more easily transformed chemically than were the larger compact and rigid blocks, the cleavage is mostly confined to the matrix and tends to wrap around the blocks, as shown in fig. 31. However, the blocks did not entirely escape deformation, and many are somewhat flattened or stretched in the direction of cleavage. Some have tapering ends or tails that extend into the cleaved matrix, as shown also in fig. 31.

Freshly broken surfaces of rocks of the Taylor Formation are green because of the abundance of green metamorphic minerals such as hornblende, chlorite, and epidote. This aspect of the rocks is well displayed in the road cuts along the Gold Lake Highway, and along State Highway 89 between Indian Falls and Crescent Mills. Elsewhere, where the rocks are not covered with moss or lichens or stained rust color with iron oxides, the surface is light green or greenish gray, tending toward white, but nowhere so light colored as are the rocks of the Sierra Buttes Formation. The lighter color of these surfaces is a result of weathering, and in the Lakes Basin District, for example, it can be easily seen that on glacially smoothed surfaces the whitened layer is at most only a small fraction of an inch thick.

Pillow lavas occur at numerous places but unfortunately no good example is close to trails or roads. The excellent pillows shown in fig. 22 are located 1,100 feet west of the Gold Lake Highway near the northwest corner of the NE 1/4, SE 1/4, Sec. 6, T. 21 N., R. 12 E. This is also about 750 feet northwest of the hill marked 6511 on the Sierra City quadrangle (15-minute series) and quite close to but east of Gray Eagle Creek. Thin-bedded andesitic mud between the pillows at this locality has been highly deformed (fig. 23), no doubt the result of the extrusion of the pillows into the sea floor of soft andesitic mud.

Remarkable pillows that contain large elongate and radially arranged amygdules of quartz, one of which is shown in fig. 59, occur on the ridge east of the Gold Lake Highway on the south side of a shallow saddle almost due east of the entrance to the Lakes Basin campground.

The top of the Taylor Formation is just a few feet west of the road at the Frazier Falls picnic grounds. This is at the Plumas-Sierra county line and at the end of pavement on the Old Gold Lake Road.

Like the formations below it, the Taylor extends on the north to Lake Almanor and on the south to the North Fork of the American River. It is much thicker than the Sierra Buttes Formation and its original volume must have been several times larger. The Taylor is thick near Lake Almanor and on Grizzly Ridge. In the Lakes Basin its thickness is about 10,900 feet, a little more than 2 miles, but 9 miles to the south on the North Fork of the Yuba River it is only 1,200 feet thick.

Such a rapid change in thickness suggests that this part of the Taylor may be a great volcano turned on edge and eroded through

Fig. 59. Pillow lava of the Taylor Formation. This remarkable pillow has elongate amygdules of quartz radially disposed. On the ridge east of the Lakes Basin Campground, about 300 feet north of the S 1/4 Cor., Sec. 5, T. 21 N., R. 12 E. For a description of the view from this point see fig. 163. Pillows are also shown in figs. 22, 23, 61, 62, and 66.

somewhere near its summit. If so, it compares favorably in size and shape to Mt. Shasta, which is also andesite. Shasta is about 10,000 feet high (it stands on a platform about 4,000 feet above sea level), the diameter of its base is about 16 miles, and its slope averages about 14.5 degrees. If the Taylor of the Lakes Basin District is a volcano with its summit near Gold Lake and one of the edges near the North Fork of the Yuba River, it would have a height of 10,900 feet, the diameter of its base would be 20 miles, and its slope angle would be 11.5 degrees.

It is not known that the Taylor Formation also thins north of the Lakes Basin as it does to the south because of complications arising out of younger faults and concealment under younger rocks beneath

Mohawk Valley, so the analogy is not perfect. Although the comparison proves nothing, the similarity to Shasta, and also, incidentally, to the existing submarine volcanoes known as seamounts that stud the present sea floor, brings to mind the idea that the Taylor Formation may consist of a series of great volcanic cones—essentially great mounds of volcanic breccia. Indeed one may carry speculation a little further. If volcanoes such as the one suggested above, nearly 11,000 feet high, were built upon the sea floor, the tops of some might well have extended above sea level so that their tops became islands. Of course, there is presently no evidence that such islands really existed; yet, as is shown in the discussions of some of the younger formations of this region, there is some reason to think that seashores may not have been very far away.

No vents through which the rocks of the Taylor Formation were brought to the surface have yet been identified. To be sure, many dikes are present in the Shoo Fly, Sierra Buttes, and Taylor formations, but they seem incompetent to have provided the enormous volume of breccia that is the Taylor Formation. I prefer to think in terms of large masses of intrusive rock representing one or more central vents. One such intrusion is present in the Plumas-Eureka State Park, and others are present both to the north and south. But since the source of the rocks of the younger Goodhue Formation is as problematic as that of the Taylor, the discussion of sources is relegated to pages farther on.

No fossils have yet been found in the Taylor so its age is not known except that it is younger than the Elwell Formation and older than the overlying Peale Formation, described next.

The Taylor, together with the underlying Elwell and Sierra Buttes formations totaling a thickness of more than 15,000 feet, no doubt represents but a small part of a vastly more extensive island arc system of volcanoes that came into existence in the sea west of North America.

Peale Formation

The Peale Formation rests on the Taylor Formation; like the Taylor it is composed largely of volcanic material but it looks very different. It, too, extends as far north as Lake Almanor and as far south as the North Fork of the American River. In the Lakes Basin District it is exposed for only a short stretch, from a little north of Frazier Falls to a point a little short of Gold Lake, a distance of only a mile and a half. At both ends it is concealed by younger rocks. It is best studied near Frazier Falls, where it can be examined on glaciated surfaces.

The base of the Peale—its contact with the Taylor Formation—is only a few feet west of the end of the pavement on the Old Gold Lake

Fig. 60. Peale Formation. This, the lowest bed of the Peale Formation, consists of blocks and smaller pieces of volcanic rock mixed with grains of the mineral calcite (calcium carbonate) that possibly were the buttonlike segments of the marine animals known as crinoids. West side of the Old Gold Lake Road at the parking area of the Frazier Falls picnic grounds.

Road at the Frazier Falls picnic grounds. This is also at the Plumas-Sierra county line. The lowest Peale rock (fig. 60) is fine volcanic breccia, the largest blocks being only a few inches in diameter. The mottled matrix consists of dark-colored scoriaceous volcanic rock fragments and white to gray calcium carbonate—the mineral calcite.

North of the parking area and east of the road there is a smoothly rounded rock surface upon which one can see the outlines of pillows (figs. 61 and 62), each with a center that looks spongy because it was once vesicular. Dark reddish or maroon-colored chert fills the spaces between pillows. East of the pillows is a series of thin beds of very fine volcanic ash. The rock is light green but weathers nearly white. On freshly broken surfaces one may see rectangular pinkish crystals of feldspar that are characteristic of much of the Peale Formation.

Walking eastward from the parking area, one crosses some of the thin-bedded white-weathering ash and next encounters dark purplish black slaty layers of ash alternating with layers that contain abundant white to gray calcite. These rocks have good cleavage, and here, as shown in fig. 30, is an excellent place to observe that original bedding and cleavage of metamorphic origin are not necessarily parallel.

A layer of hard volcanic sandstone about 1.5 feet thick is inter-

Fig. 61. Peale Formation. Reddish pillow lava with pieces of maroon radiolarian chert between the pillows. About 600 feet north of the parking area at the Frazier Falls picnic grounds on the Old Gold Lake Road. The scale is given by the seated figure in the lower left center.

Fig. 62. Peale Formation. A close view of the pillows shown in fig. 61. The pillows have been deformed into lenticular shapes and their margins are crenulated. The centers of the pillows weather more readily than the margins because they were vesicular. The scale is given by the pencil at the center of the picture.

Fig. 63. Peale Formation. Gray limestone composed of the small buttonlike segments of the marine animals called crinoids, with lesser amounts of small black particles of volcanic ash. East of the parking area at the Frazier Falls picnic grounds, on the Old Gold Lake Road. The rocks are folded into a small syncline. The hammer sets the scale.

bedded with the slaty rocks. It consists of graded beds, beds disturbed by the deposition of a subsequent layer, and, in places, of ripple-marked layers that are cross-bedded (fig. 14). Here one may easily see that these features show that the top faces to the east.

On this east-facing slope a cream-colored limestone consists of the buttonlike segments of the disarticulated marine organisms known as crinoids, mixed with a small amount of fragments of dark volcanic scoria. The limestone is present in numerous places owing to small folds and faults (fig. 63).

A short half mile south of the parking area on the east side of the road an extensive clear area of glacially smoothed rock is entirely of the fine-grained white-weathering ash. Here one can see very fine laminations, intricate faulting, and disturbances due to the slumping of layers already deposited, and even the downward intrusion of one layer of ash into another, as shown in fig. 64. Much of the obvious deformation occurred while the sediment was still soft, long before the Nevadan Orogeny and metamorphism.

East of Frazier Creek, and about a quarter mile south of the bridge on the trail from the parking area to the overlook of the falls (in

Fig. 64. Peale Formation. Downward intrusion of the light-colored volcanic ash into dark volcanic ash. This intrusion probably occurred at the time of deposition rather than as a consequence of orogeny. East of Old Gold Lake Road, 0.5 mile south of the Frazier Falls picnic grounds.

the NW 1/4, Sec. 9), near a pond, which is usually dry late in the summer, the radiolarian chert, volcanic sand, and a bed of conglomerate that contains both rounded and angular fragments are exposed. Some of the fragments of the conglomerate are of pink granite that appear to have been metamorphosed before they became incorporated into the Peale. The first ridge northeast of the pond is of lighter-colored rock, shown in fig. 65, that seems to consist of flattened fragments of pumice.

Farther east as far as the top of the Peale, which is in the principal ravine east of Frazier Creek, the rocks are mostly dark-colored volcanic ash with various amounts of calcium carbonate, and some beds of impure limestone.

The conditions under which the Peale Formation originated are not clear. Near-surface conditions are indicated by the conglomerate, the volcanic rock fragments, the scoria, and the pumice. Shallow water might be indicated by the strong currents required to produce ripple marks, but strong currents are known to occur in the deep ocean. A shallow-water environment for their dwelling place is indicated by the crinoids of the limestone, but their remains could

Fig. 65. Peale Formation. A rock once composed mostly of fragments of pumice, now compressed and deformed around the rigid nonporous rock fragment at the center of the picture. Although this rock was submarine in origin it closely resembles some of the welded tuff of subaerial origin that comprises most of the Delleker Formation, described in Chapter 6.

have accumulated in shallow water and then slid into deep water. However, proof that such sliding occurred is lacking. The radiolarian chert seems to indicate deep water.

Although organic remains are abundant in the Peale near Frazier Falls, none have been found that could reveal its age. Fossils collected from the Peale at the junction of Cascade and Little Grizzly creeks early in this century, and collected again in the 1970s, indicate that it was deposited in early Mississippian time. The Peale Formation near Frazier Falls is about 2,600 feet thick.

Arlington Formation

The Arlington Formation is named for its occurrence on Arlington Heights at the north end of Grizzly Ridge, where it is next above the Peale Formation. Like the older formations, its top faces to the east. The formation is 8,000 feet thick and consists of water-laid volcanic sand and mud, with lesser amounts of volcanic breccia. With the exception of the black mud, which is now slate, the rocks are much like the lower part of the Taylor Formation.

North of Grizzly Ridge the Arlington continues as far as Lake Almanor, and equivalent rocks are present also 60 miles farther northwest in Shasta County. It is present but not well exposed along State Highway 89 in Secs. 25 and 26, T. 26 N., R. 9 E., between Indian Falls and the road to Taylorsville. There it is a greenish gray, sandy-appearing rock.

Farther west the Arlington underlies a large area between the ridge east of Butt Valley Reservoir and Yellow Creek, a distance of 6 miles. The rocks of that area are dominantly andesitic volcanic ash and volcanic sand, alternating with layers of black slate and lava flows of andesite and basalt. Only a little is volcanic breccia. Complexly folded and faulted, their thickness is estimated to be 7,000 feet. State Highway 70 crosses the belt of Arlington between Rich Bar and Halsted Flat, a distance of about 2 miles.

In the Lakes Basin District, rocks that are probably part of the Arlington Formation are visible on the steep slope, almost a cliff, between Mills Peak and Frazier Creek. The slope is visible from the Old Gold Lake Road and from the vicinity of Frazier Falls. The rocks of the lower part of the slope are andesitic sandstone with fossils that are the segments of crinoids and angular fragments of the shells of brachiopods, none of which have yet been found adequately preserved to provide an age. Higher on the slope the rocks are all andesite breccia that contains blocks of a dark-colored mafic plutonic rock similar to that on Eureka Peak. The thickness is about 4,000 feet. Black slate is not present, nor is limestone or lava flows, but the differences between the rocks here and farther north should not be surprising as the intervening distance is not less than 25 miles. An important feature common to all is the presence of original hornblende, which serves to distinguish the Arlington rocks from both older and younger volcanic rocks.

On Grizzly Ridge the Arlington is limited on the east by the Grizzly Mountain Thrust Fault, which divides the eastern belt of metamorphic rocks into two parts in which the sequence of rocks is repeated. If the rocks between Frazier Creek and Mills Peak are really of the Arlington Formation as I believe them to be, their eastern border, only a short distance west of Mills Peak lookout, might very well be the Grizzly Mountain Thrust Fault. At that place the rocks do change abruptly from andesite to the much darker colored basalt of the Goodhue Formation, described next.

On Grizzly Ridge the Arlington contains some rocks identical to those of both the Goodhue and Reeve formations east of the Grizzly Mountain Thrust Fault, which are therefore thought to be its equivalent, at least in part.

The age of the Arlington is established by fossils from the rocks around Butt Valley Reservoir as ranging from Lower Permian to Upper Triassic.

The thickness of the Arlington between Frazier Creek and Mills Peak is perhaps sufficient for the accumulating volcanic pile to have been built above sea level. The organisms that secreted the fossil shells certainly lived in water no more than a few hundred feet deep. Broken but not rounded as would be expected had they been rolled by waves for even a short time, the shell fragments are, perhaps, good indicators that the shells were transported by sea-floor sliding from their place of origin into deeper water.

The shallow water in which the organisms lived could not have been along the shore of the continent because that was still more than 200 miles farther east. It must have been, then, either around a volcanic island or over a volcanic platform or shoal built to near sea level. Firm evidence that an island existed is lacking, as is any indication of the form of a volcanic cone as there is in the case of the Taylor Formation. However, in the light of the continuity of the Arlington Formation, there might well have been a chain of volcanic islands.

It should be kept in mind, however, that the volcanoes of the Arlington Formation are in no way related to those of the Taylor Formation, because the time between the end of eruption of the Taylor volcanic rocks and the beginning of those of the Arlington may have been as long as 100 million years.

Goodhue Formation

The Goodhue Formation was named for the Goodhue homestead on Wards Creek southeast of Genesee, where it rests on the Peale Formation. It continues for some miles north of Genesee Valley but is absent from Keddie Ridge. To the south it is present on Grizzly Ridge as far as Cascade Creek, where it ends against a pluton of granitic rock. It occurs again at Clio, and, between Mills Peak and the Mohawk Saddle, whence it continues as far south as the Middle Fork of the Yuba River. At Mills Peak it seems to be bounded on the west by the Grizzly Mountain Thrust Fault about 1,500 feet southwest of the lookout tower.

Pillows of basalt lava that have concentric zones of amygdules (fig. 66) and breccia made of amygdaloidal pieces of pillows (fig. 67) can be seen in the crags west of the Mills Peak lookout tower. Well-graded beds of volcanic sand are present east of the lookout tower at the lip of the cliffs by the road.

On the upper side of the road about a quarter mile from the lookout tower there is a pit or quarry in radiolarian chert, which, when

Fig. 66. Goodhue Formation. Pillow lava of basaltic composition, with concentric zones of amygdules. About 300 feet west of and below the lookout tower at Mills Peak.

freshly broken, is about the color of milk chocolate. One can walk entirely around the chert and not discover any continuation of it away from the quarry. It may be that it is a huge block that slid on the sea floor into place among the volcanic rocks. Other smaller blocks of similar chert occur in volcanic breccia on the ridge about a mile southeast of the lookout tower.

Between Mills Peak and Mohawk Saddle, layers of basalt breccia alternate with layers of volcanic sand. The rocks are very dark colored, nearly black in contrast to the light color of the andesite of the Taylor, Arlington, and Reeve formations. White phenocrysts that once were feldspar and others of glassy green augite are conspicuous. Except for the small amount of chert, the Goodhue is volcanic. To the north the Goodhue extends beyond Taylorsville and in the south beyond the North Fork of the Yuba River. Its thickness of 5,100 feet near Mills Peak is probably greater than it is elsewhere. This is the best place to examine the Goodhue, although large blocks of the black breccia are easily reached along Little Grizzly Creek in Secs. 2 and 35 between the now inactive Walker Mine and Genesee Valley.

No fossils have been found in the Goodhue Formation so its age

Fig. 67. Goodhue Formation. Volcanic breccia composed of amygdaloidal fragments of pillows. This is the bed next above the pillow lava shown in fig. 66. It is overlain in turn by beds that are successively finer grained as far as the coarse sand that is exposed in the cliffs 1,000 feet southeast of the lookout tower.

is not known directly. It is older than the overlying Reeve Formation of Permian age, and it is younger than the Peale, which is directly beneath it in Genesee Valley. Since the Peale is Mississippian, the Goodhue is probably Permian.

Reeve Formation

The Reeve Formation is next younger than the Goodhue Formation on which it rests and like the Goodhue it is almost exclusively volcanic. It also faces to the east, inclined about 60 degrees, and is known from Lake Almanor to the Middle Fork of the Yuba River. Its contact with the Goodhue is not easily located because one grades into the other. The gradational contact between them passes through the Mohawk Saddle and extends down the mountain face to the north.

Nearest to rocks that are certainly Goodhue, the Reeve is a fine-grained bluish gray metamorphosed volcanic ash and breccia. Next above that (to the east) there is an impure limestone with obscure traces of fossils, and above the limestone there are well-bedded fine-

Fig. 68. Reeve Formation. A block of bluish gray lava from a volcanic breccia. The square-to-rectangular white crystals (phenocrysts) of feldspar characterize the Reeve Formation. On the railroad 0.5 mile west of the "C" Road crossing at Clio.

grained metasedimentary rocks with some layers of andesite ash and breccia. Only the bluish gray rock looks very much like the Reeve of the Taylorsville District, where it was first named and described.

Rocks of the Goodhue and Reeve formations are also present on the north side of the Mohawk Valley, north of Clio, along the railway tracks. There the lowest part of the Reeve consists of a volcanic breccia of bluish gray rock with one-fourth-inch-square phenocrysts of white feldspar (figs. 68 and 69). Some of the breccia has a calcareous matrix that contains fragments of the marine organisms brachiopods and crinoids. Fine-grained, well-bedded sedimentary rocks with a few beds of volcanic breccia, all exactly like the rocks on the slope north of the Mohawk Saddle, are present along the railway from the road crossing eastward to the end of the great Willow Creek trestle (fig. 70).

The Reeve is also known at several places between Clio and the now inactive Walker Mine. From there it is continuous across the Taylorsville District and for many miles northwest of Genesee. To the south it extends beyond the North Fork of the Yuba River.

The Reeve is about 2,000 feet thick at the Mohawk Saddle and

Fig. 69. Reeve Formation. Volcanic breccia. Some but not all of the fragments contain the typical white phenocrysts of feldspar shown in fig. 68. The location is the same as in fig. 68.

3,000 feet thick on Keddie Ridge. Its age, based on fossils found near the Walker Mine and at Clio, is Middle to Late Permian.

The nature of the environment in which the Reeve was deposited seems to have been the same as that of the Arlington Formation, to which it is partly equivalent. Although there is no evidence of volcanic islands, the organisms represented by the fossils certainly lived in shallow water. The fragmental condition of their remains, like those of the Arlington Formation, probably means that they, too, either lived on a rocky bottom where the shells became broken by waves, or that shells and rocks together slid into deeper water, the shells becoming broken in the process. That the shell fragments are not rounded by wear, as would be expected if they were rolled around by waves, makes the idea of sliding more attractive. Later, the fine-grained, well-bedded sediments were deposited on the breccia.

Rocks of Triassic Age

Rocks of Triassic age have long been known in the vicinity of Butt Valley Reservoir and in Genesee Valley. Two formations—the Swear-

Fig. 70. Reeve Formation. Thinly bedded, fine-grained volcanic ash. The beds are steeply inclined to the right. On the railroad west of the "C" Road crossing at Clio.

inger, of sooty black slate, and the Hosselkus, which is limestone—occur in a narrow strip only 3 miles long, extending north from Genesee Valley in Sec. 10, T. 25 N., R. 11 E. The slate, so rich in carbon of organic origin that it will blacken one's fingers, contains abundant impressions of pectenlike shells one to two inches in diameter. The limestone also contains abundant shells, especially those of an ammonite less than an inch in diameter. These and other fossils clearly indicate a Triassic age. The two formations together are only about 1,000 feet thick and are remarkable in this region for being devoid of volcanic debris, although that is, no doubt, only a local phenomenon.

The oldest of the two formations, the Hosselkus, is believed to rest unconformably on the Reeve Formation, but there is no place where the unconformity can be examined as can the one at Wades Lake.

Farther west, as already noted, rocks of Triassic age are included in the Arlington Formation, which is mostly volcanic. Fossils in limestone on the ridge east of Butt Valley Reservoir, and also near Yellow Creek in Secs. 17 and 18, T. 26 N., R. 7 E., are Triassic in age. Other rocks of Triassic age are unknown in the northern Sierra Nevada but they do occur in Shasta County 60 miles north, in the southern Sierra Nevada, and in central western Nevada. At all of these places the rocks are marine and largely volcanic.

The dominantly volcanic rocks of the Arlington, Goodhue, and Reeve formations along with others of Triassic age, some 19,000 feet or 3.5 miles thick, are a part of a thick and extensive sequence that comprise another volcanic arc system that came into existence in the ocean far west of the shore of North America (see fig. 39). This arc, analogous to that composed of the Sierra Buttes, Elwell, and Taylor formations, was constructed above that one, as shown in fig. 72.

Rocks of Jurassic Age

The Jurassic rocks are best known on Mt. Jura[1] at Taylorsville, and farther east where they occupy a strip as much as five miles wide that includes Wheeler Peak and Kettle Rock. The strip extends north as far as State Highway 36 and south to the Walker Mine. East of Genesee the Jurassic rocks, about 15,000 feet thick, are folded into the great syncline shown in figs. 73 and 74. Fossils of marine organisms are abundant. For example, a rust red sandstone on the hill slope a quarter mile north of the fairground at Taylorsville contains a wealth of impressions of broken shells.

Most of the Jurassic rocks are volcanic sediments, some of which are quite similar to those of the Taylor Formation. However, a scarcity of chert and an abundance of sandstone, conglomerate, and fossils of organisms that lived in shallow water suggests that deposition occurred in water shallower than that of previous times. Such shallow water deposits suggest in turn the presence of volcanic islands or shoals of volcanic origin for it is quite certain that the continental shoreline was still far to the east.

The rocks of Mt. Jura and the region east of it have been divided into fourteen formations. I have refrained from describing them because the region is so well covered with soil and vegetation that directions to any particular place cannot easily be given, and the rocks are not well exposed.

Near the Walker Mine the Jurassic rocks terminate against intrusive granitic rock. Other similar volcanic rocks are present farther south along the Lake Davis Road north of Portola; along the railway track and the Middle Fork of the Feather River east of Clio and Willow Creek; and east of the Mohawk Saddle. No fossils have been found at any of these places, but because they are in line with the Jurassic rocks at the Walker Mine and are east of the Reeve Formation they are possibly Jurassic. Other fossiliferous Jurassic volcanic rocks of similar character occur along Interstate Highway 80 near Cisco Grove, west of

1. Mt. Jura was named because of the presence there of fossiliferous rocks of Jurassic age. The Jurassic period takes its name from the Jura Mountains of Switzerland.

the south end of Lake Tahoe, and in Shasta County. As shown farther on, rocks of Jurassic age occur as far west as the Sacramento Valley.

ROCKS OF THE WESTERN BELT

Smartville Complex

The Smartville Complex underlies a belt as much as 25 miles wide that continues to the west beneath the younger sedimentary rocks of the Sacramento Valley. From about 12 miles north of Oroville it extends beyond Auburn, but its southern limit is not yet established. It is at least 4 miles thick and perhaps half again that much.

The complex differs from the rocks of the eastern belt in that it does not consist of formations deposited one on the other in time sequence. It is simply a group of physically related rocks to be described and explained. The word complex has no more specific meaning than that. The complex consists of three layers; a lower layer of gabbro and related plutonic igneous rock, metamorphosed, of course, as are all the rocks of the complex; a middle layer of basaltic igneous rocks, in large part dikes; and an upper layer of lava flows, pillow lava, brecciated lava, and volcanic breccia. The rocks of all three layers are essentially basaltic and andesitic in composition—that is, mafic and intermediate—although silicic rocks are present in very minor amounts. Such rocks in kind and in sequence clearly resemble, but are not identical to, what is believed to be the nature of the earth's oceanic crust.

At the midocean ridges, a great system of cracks that encircle the earth, mafic magma is extruded to the sea floor as lava flows and pillow lava, and no doubt also as brecciated forms of volcanic rock. This igneous nature of the sea floor has been inferred from many kinds of observations, not least from direct observation from deep-diving submarines. The magma extruded upward from a deep source in the mantle fills a fracture, thereby producing not only lava but also a dike. Since the ocean floor divides at the midocean ridge, each side moving away from the ridge, repeated intrusion constructs a layer of dikes below the lava. At the same time, part of the magma below the layer of dikes crystallizes to coarse-grained plutonic rock, thus providing a third layer to the ocean crust. One would not expect the layers to be sharply separated from each other; there should be interfingering. In addition, above the lava is another layer, one of sediment that consists of the insoluble remains of organisms that lived and died in the sea, windblown dust from land above sea level, and even meteoritic dust from outer space. Much of the sediment is siliceous and becomes radiolarian chert.

This sequence of layers of the ocean crust is not merely inferred; it is observed in bodies of rock called ophiolites, which occur in many places around the world. Furthermore, it has recently been confirmed in part by drilling into the sea floor from the exploratory ship *Glomar Challenger.*

Thus the Smartville Complex is an ophiolite, or a body of ophiolite, but it is not complete for it lacks the sedimentary layer with chert, and perhaps it has more brecciated lava than is common in other ophiolites. Even though incomplete and not entirely representative, it can hardly be thought of as anything but a huge slab of oceanic crust that has been brought from an original deep level to a higher position into what is now continental crust.

The means by which oceanic crust attains a higher level is called obduction, the opposite of subduction. Instead of continuing to slip beneath an opposed plate along a subduction zone, a part of the oceanic crust apparently slides across the subduction zone and the edge of the opposed plate on which it comes to rest. Many explanations have been offered for the mechanisms of such events, but none seems to be universally acceptable. The problem will no doubt be resolved at some time in the future.

Since the Smartville Complex is oceanic crust, it is possible that beneath the lower layer of plutonic rock there is a still-deeper and not presently exposed layer of ultramafic rock that was part of the earth's mantle. Such a layer is present in ophiolites elsewhere in the world.

The age of the Smartville Complex is clearly established as Jurassic by radiometric dating methods. At least three determinations yield ages of about 160 million years. It is, of course, older than the overlying Monte de Oro Formation, whose age, based on fossils, is late Jurassic.

Monte de Oro Formation

This unit is known only in an area of less than a square mile about 3 miles northeast of Oroville, mostly in Sec. 33, T. 20 N., R. 4 E. It consists of conglomerate, sandstone, and dark-colored slate of volcanic origin. It contains a variety of the shells of marine animals, mostly broken, and abundant remains of a large variety of land plants. These features plus cross beds and filled channels indicate shallow-water conditions. The age, based on both plant and animal remains is late Jurassic. Like the Jurassic rocks of the Taylorsville District, these indicate that the sea was shallowing, and that the construction of the Nevadan geosyncline was nearing an end.

ROCKS OF THE CENTRAL BELT

The rocks of the central belt are separated from those of the eastern belt by the Feather River Serpentinite, described farther on. Like those of the eastern belt, the rocks are dominantly volcanic, and large volumes of quartz-rich sediment like the quartzite of the Shoo Fly Formation are lacking. Probably such land-derived sediment was never transported so far west.

The volcanic rocks are breccia, sand, and mud of andesitic and basaltic composition much like those of the Taylor Formation in the Lakes Basin District. Lava flows, including pillow lava of basalt and andesite, are also abundant. Silicic volcanic rocks, also well represented, were probably deposited as ash. A very important part of the rocks of the central belt consists of chert associated with fine-grained sediment, both mud and ash. Serpentinite, also an abundant rock type, is discussed farther on in connection with the Feather River Serpentinite.

The best place to examine the rocks of the central belt is along State Highway 70 from its junction with the Cherokee Road north of Oroville to Jarbo Gap and from the Rock Creek Dam to Rich Bar. Central-belt rocks are also accessible along State Highway 49 from a few miles west of Camptonville to Goodyears Bar, and along the Marysville–La Porte–Quincy Road from Challenge to the crossing of the South Fork of the Feather River near Onion Valley. An excellent exposure of pillow lava is at the dam of Little Grass Valley Reservoir.

Because the rocks of the central belt suffered greater disruption by deformation than did those of the eastern belt it has been difficult to establish the order of succession of the rocks and, therefore, their relative ages. In fact, deformation has been so disruptive, especially in the rocks originally composed of chert, mud, and ash, that over much of the region the original layering has been destroyed. Accordingly, large volumes of rock now consist of a jumble of blocks of harder or stronger rock in a matrix of softer, weaker rock. Such masses of rock are called melange, which is not a varietal rock name but one applied only to such deformed bodies of rock. A zone of melange is exposed along State Highway 70 east of the crossing of Oroville Reservoir.

Fossils are rare and seem to be only in blocks of limestone in melange. Their Permian and Pennsylvanian age applies only to the limestone that contains them. The matrix that surrounds the blocks remains undated; it could be the same but it could be either younger or older. Blocks of fossiliferous limestone are present along the west shore of Oroville Reservoir north of State Highway 70, especially near the center of Sec. 17, T. 21 N., R. 4 E. The age of some volcanic rocks determined by radiometric methods, a matter explained in Chapter 5, is Jurassic.

It is difficult to establish formations in such highly deformed rock with few or no meaningful fossils. Although much progress has been made in recent years in understanding the rocks of the central belt, it has not been possible to equate any of the rocks, other than those of Jurassic age, to any of the formations so clearly defined in the eastern belt. On the other hand, some of the rocks of the central belt, especially the basaltic and andesitic volcanic rocks of Jurassic age, are quite similar to parts of the Smartville Complex of the western belt. They may once have been part of that great slab of ocean crust.

FEATHER RIVER SERPENTINITE

The fourth belt of the metamorphic rock of the northern Sierra Nevada is the Feather River Serpentinite, a body of massive rock that lies between the eastern and central belts (figs. 40 and 41). Composed mostly of the rock serpentinite with lesser amounts of related rock, the origin of which is discussed farther on, it is a large slab of what was quite possibly a part of the earth's mantle. Three miles wide (4 miles by road), where it is crossed by the east branch of the Feather River and State Highway 70 between Virgilia and Rich Bar, it extends to the north across the North Fork, a distance of 6 miles. To the south it continues at the same width across the Middle Fork and Onion Valley Creek, a distance of about 20 miles. There it narrows and at the South Fork of the Feather River is less than a mile wide. From there south it continues mostly as a narrow body, partly concealed beneath younger rocks, and perhaps discontinuously at the surface, to its southern end south of the Middle Fork of the American River. Its total length is about 75 miles and its volume is at least 100 cubic miles and perhaps several times that.

Both sides (contacts) of the belt are faults. That on the east is the Melones Fault and that on the west is the Rich Bar Fault. A narrow strip of rock metamorphosed at a temperature higher than is prevalent in the northern Sierra Nevada lies along the western edge. Study with the microscope has revealed that the rock was intensely deformed before the alteration to serpentine minerals took place. A radiometric date indicates that the rock may be nearly 400 million years old, about as old as the Shoo Fly Formation.

The ultramafic composition of the body, its early deformation, the presence of the high-temperature metamorphic rocks on the western contact, and the fault nature of its boundaries have been taken to indicate that the body is a piece of the earth's mantle brought up into the geosyncline from great depth. It is not an ophiolite because it is unlike the ocean crust of the Smartville Complex; neither can its position be related to either subduction or obduction. However, its position may be connected with an older concealed subduction zone associated with the development of the island arc rocks of the Sierra Buttes, Elwell, and Taylor formations. The matter is discussed further in the next chapter.

ABOUT SERPENTINITE AND RELATED ROCKS

The rock serpentinite[2] originates by the alteration of ultramafic igneous or metamorphic rock. The original minerals were pyroxene and olivine, which are hard and vitreous appearing. The secondary or derived minerals are those of the serpentine group, which are soft, flaky or fibrous, and usually too fine grained to be visible to the naked eye. (Most asbestos is a member of the group.) The rock serpentinite is also soft like the minerals that comprise it and can be easily scratched by a knife or a piece of hard rock. Its color is varied, being mostly dark green to black on freshly broken surfaces. On naturally weathered surfaces it is commonly brown, brownish yellow, bluish, or dull violet. The exposed surfaces are commonly glossy and smooth, slick, and oily or greasy looking, an appearance owing to shearing of the rock, a consequence of deformation. Much serpentinite is so intensely sheared that the smooth, slick surfaces are so closely spaced as to make it virtually impossible to break a specimen to show an unsheared surface. An end product of such shearing is a rock composed of nothing but small smooth and shiny scalelike particles. The soil derived from serpentinite is notably thin and poor in nutrients; hence the vegetative cover is scanty and poor in quality.

Serpentinite, the official California state rock,[3] is common in the folded mountain systems of the world and is particularly abundant in northern California. It is usually associated with the submarine volcanic rocks of geosynclines and with the processes of subduction and obduction, and the rock of the upper part of the earth's mantle is

2. The name serpentinite, often pronounced in an affected manner with accent on the second syllable, is derived from the older name serpentine, accented on the first syllable. The name serpentine was once applied both to the rock and to a group of minerals that are the principal constituents of the rock. The reason for the change was to make a clear distinction between the rock and the minerals.

3. In September 1965, then Governor Edmund G. Brown, the elder, declared serpentine to be the official state rock and gold to be the official state mineral.

thought to be composed of serpentinite's parent rocks.[4] Serpentinite occurs as enormous masses like the Feather River Serpentinite described above, but also as layers, lenses, or beds whose dimensions are measured merely in feet.

The parent rocks of serpentinite were in many instances igneous, and plutonic igneous rocks such as gabbro are common associates. Such is the case, for example, with the Feather River Serpentinite, which may be inferred, therefore, to be at least in part igneous. The same is true of many other serpentinites of the northern Sierra Nevada.

Serpentinite is not very abundant among the metamorphic rocks of the eastern belt. A small body is present in Big Jamison Creek Canyon; it is crossed by the Johnsville–La Porte Road about one and a half miles upstream from the entrance to the Plumas-Eureka State Park Campground. Several small bodies occur on the west slope of Sierra Buttes, and another is present northwest of Grizzly Peak on Grizzly Ridge. Other bodies occur some miles south and west of Greenville.

Serpentinite is abundant in the central belt. Some masses there, as shown in fig. 40, are more than a mile wide and 6 to 8 miles long, but many are quite small. An intensely deformed mass of the associated coarse-grained plutonic igneous rock gabbro can be seen along State Highway 70 a mile south of the twin bridges near Pulga (a dangerous place to stop). This is close to the Camel Peak Fault (see fig. 40).

Emplacement by obduction and igneous intrusion are not the only means by which ultramafic bodies that came to be serpentinized arrived at the positions as seen today. Once serpentinized, the rock, being soft and slippery, is subject to easy movement in the solid state. Folding and faulting can divide large bodies into smaller ones, especially under the disruptive conditions that prevailed in the central belt of metamorphic rocks. Serpentinite is known to have been injected along faults to form dikelike bodies. It could be intruded between the layers of lava and fragmented volcanic sediment to appear as sill-like or bedlike bodies. Some serpentinite is truly sedimentary, having been eroded from exposed bodies of serpentinite that was emplaced in other ways, or by having been extruded onto the sea floor by breaking through overlying layers of rock, a process called diapirism. Landsliding and, no doubt, seafloor slumping are important in the emplacement of sedimentary serpentinite. None of these processes has been demonstrated for any bodies in the northern Sierra Nevada but most have been demonstrated elsewhere.

4. The parent rocks, composed of various relative amounts of pyroxene and olivine, are named pyroxenite, dunite, and peridotite. The latter name is derived from peridot, an old name for olivine, which is still in use for the clear green gem-quality olivine.

All of the serpentinite bodies of the northern Sierra Nevada were deformed and metamorphosed during the Nevadan Orogeny, the subject of Chapter 4. That event transformed some of the serpentinite into soapstone, as at Soapstone Hill in Sec. 22, T. 23 N., R. 6 E., about 8 miles southwest of Bucks Lake. But the strong deformation of the Feather River Serpentinite and of the gabbro south of Pulga probably occurred at an earlier time while the rock was still in the mantle or while it was being transferred into the crust.

Serpentinite can be easily reached for observation along State Highway 70 near Jarbo Gap on both sides of the gap, between Jarbo Gap and Pulga, and between Virgilia and Rich Bar. Serpentinite is excellently exposed almost without vegetative cover along the Johnsville–La Porte Road at Whiskey Diggings north of Gibsonville townsite.

OTHER METAMORPHOSED IGNEOUS INTRUSIONS

The immense volume of volcanic rock in the central and eastern belts is not the kind erupted from the fissures of midocean ridges, where ocean crust originates. Instead it is like the rock of the great chains of volcanoes associated with the present subduction zones, such as those of the Aleutian Island or the East Indies chain. The volcanoes of those regions erupt through localized or central vents that are roughly circular in plan and cylindrical to some depth. When the eruptive cycle is completed the vent remains as an intrusion of igneous rock called a plug.

Many volcanic centers seem to be located at the intersections of deep-seated fractures or fracture systems, so the cylinder at the surface may be irregular at depth, being bounded by former fractures, or it may pass into one or two or many dikes that fill fractures along which magma rose from greater depth.

The recognition of ancient vents may be difficult. What may be found on the present erosion surface, a random slice through the deformed volcanic pile, depends in part on the depth of erosion. The form may not be that of a cylinder, but that of a lens, or it may be highly irregular. The rock may not be volcanic in character but may be coarser grained, perhaps even plutonic because it cooled slowly at depth in the vent. A vent may have been disconnected from its eruptive products by faulting. All things considered, it is perhaps remarkable that any vents should be recognized among old volcanic rocks. Yet there may be some among the rocks of the eastern belt.

In my discussion of the Sierra Buttes Formation I mentioned that two areas had been suggested as source vents. The rock at both places is certainly of volcanic or near-surface origin, but at neither place has the form of a vent been demonstrated. Several possible centers of eruption of mafic and intermediate (basaltic and andesitic) rocks, such as those of the Taylor, Arlington, and Goodhue formations, are known.

One such intrusion is present at Johnsville, but so much of it is covered by younger rocks that its size and shape remain somewhat uncertain. It is composed of both metagabbro and metapyroxenite. The gabbro is a granitelike plutonic rock that does not contain quartz, and about half of which is dark minerals (see fig. 18). It is green because of the minerals produced by the metamorphism. The pyroxenite is an ultramafic rock composed of the dark green mineral augite, a member of the pyroxene family, with only traces of feldspar. It is also coarse grained (see fig. 78).

The metagabbro forms most of the east slope and some of the north slope of Eureka Peak, and a small, separate, oval-shaped body of similar rock extends from the ridge west of Eureka Lake into the Deer Creek Basin. Gabbro is also present at the Johnsville swimming hole, at the Jamison Mine, at the junction of the trails to Smith and Jamison lakes, and on the north face of Mt. Washington. The metapyroxenite forms the higher part of the northeast slope of Eureka Peak, where it can be easily reached by walking around the east side of Eureka Lake. It is also present on the point of the hill east of the Jamison Mine buildings. All of these occurrences could be parts of a single roughly circular pluton or plug about 2 miles in diameter. The pluton intrudes the Shoo Fly and Sierra Buttes formations; it is not known that it intrudes the Elwell and Taylor formations because of concealment by younger rocks.

Fragments of pyroxenite have not been found in either the Taylor, Goodhue, or Arlington formations, but fragments of metagabbro are present in the Arlington Formation between Frazier Creek and Mills Peak. Although not identical to the rock of the intrusion at Johnsville, they are sufficiently similar to indicate that the intrusion is a possible source of the andesite of the Arlington. On the other hand, the fact that gabbro fragments have not been found in the Taylor Formation proves nothing. However, this intrusion cannot have been a source for both because of the age relationships. If it had been a source for Taylor rocks, it would have been long extinct before the Goodhue was erupted, and if it were the source for the Arlington, it would have come into existence only after both the Taylor and Peale formations were deposited. Because the age of the intrusion is unknown, no further conclusions are justifiable.

Many dikes are present around Johnsville, associated with the metagabbro and metapyroxenite. They are not very much like the volcanic rocks and seem unlikely to have been sources of them.

Another but smaller metagabbro pluton similar to that at Johnsville is present at the Four Hills Mine and Upper Spencer Lake about 4 miles south of Johnsville. It is entirely within the Shoo Fly Formation, but dikes whose origin appears to be connected with the pluton extend through the Taylor Formation and intrude the Peale Formation near Frazier Falls. This fact indicates that the dikes and pluton are younger than the Peale. It remains then as a possible source for volcanic rocks of the Arlington, Goodhue, or Reeve formations or even of rocks of Jurassic age. A third, small body of similar metagabbro present near Dugan Pond north of Sierra Buttes provides no further clues.

Another intrusion of metamorphosed igneous rock is present in Long Valley, both north and south of State Highway 70. Unfortunately it is so intensely weathered that its form and character are difficult to discern, but it is porphyritic with phenocrysts of quartz and feldspar. It is too silicic to have been a source for the Goodhue Formation, but it cannot be ruled out as a possible source for the Taylor or Arlington formations or of rocks of Jurassic age.

A sill several hundred feet thick that was intruded near the top of the Goodhue Formation extends northward down the face of the mountain from the Mohawk Saddle. The rock, pictured in fig. 16, is characterized by very abundant conspicuous tabular phenocrysts of altered feldspar as much as an inch long and a quarter inch thick embedded in a dark green matrix.

The matrix of the rock weathers more easily than the feldspar crystals, which are thereby freed from the rock. They can be gathered on the Mohawk logging road in Sec. 11, T. 21 N., R. 12 E. The same sill is also present north of the Mohawk Valley west of Clio along the railway and along the ravine followed by the "C" road, which runs north from Clio to State Highway 70. Loose feldspar crystals can be gathered at both places. At Clio the sill is between the Goodhue and Reeve formations.

Rock virtually identical to that of the sill occurs at several places between the Gold Lake Highway and the A-Tree Road near the center of Sec. 33, T. 22 N., R. 11 E. All the occurrences seem to be parts of a dike that may have been the feeder for the sill. If that is the case, the dike crosses part of the Shoo Fly Formation and all of the Sierra Buttes, Elwell, Taylor, Peale, and Goodhue formations. Its age is unknown; it could be as young as Jurassic.

THICKNESS OF THE NEVADAN GEOSYNCLINE AND THE DEPTH OF THE SEA

The total thickness of the layered rocks of the northern Sierra Nevada is still undetermined but must be enormous. The thickness of the Jurassic Smartville Complex of the western belt is at least 20,000 feet and may be as much as 30,000 feet. The thickness of the rock in the central belt cannot even be estimated, but it, also, must have amounted to several miles. The thickness of rock in the eastern belt is fairly well known and is very large. Although the thickness of the Shoo Fly Formation is not known, it can hardly be less than 10,000 feet and is perhaps very much larger. That of the Sierra Buttes through Reeve formations is 29,000 feet, to which must be added another 15,000 feet of the formations of Triassic and Jurassic age of the Taylorsville District. The total is more than 54,000 feet, or about 10 miles. That is far more than the greatest known depth of the sea, which is 6.5 miles, and far more than the sea's average depth of about 12,500 feet, which is a little more than 2 miles.

It is not necessary to think that at the beginning there was a sea 10 miles deep that simply filled up with sedimentary and volcanic rock. In fact, there is evidence to the contrary. The Shoo Fly and Peale formations contain limestone that surely was not deposited in such deep water. The depth must have been closer to 12,000 feet or possibly less. Volcanoes of the Taylor and Goodhue formations, and possibly also those of the Peale and Reeve formations, may have risen to or above sea level to form platforms or islands.

The resolution of the problem lies in a behavior of the earth's crust not previously mentioned in this discussion. It is that a load added to the crust, or, conversely, removed from it, causes a depression or a rise of the surface, loaded or unloaded. The process is called isostatic adjustment.

The matter first came to light around 1850 after a discrepancy was found in the length of a line surveyed on the lowland plain of India extending to the south from the foot of the Himalaya Mountains. The discrepancy was between the length of the line as determined by surface measurements and that determined by astronomical observations.

The surface measurements involved the use of plumb bobs to establish the vertical direction—the direction along which the force of gravity acts. This would be toward the center of the earth if the earth were a smooth sphere of uniform internal composition. But the earth is not like that and a plumb bob is deflected by nearby masses of rock. So it was thought that the discrepancy in the survey might be due to the attraction of the Himalaya Mountains, which should draw the plumb bob toward them, more strongly at stations nearby, and less so at stations farther away.

When the effect of the mountains on the plumb bob was calculated, the surprising result was that the actual discrepancy was only a tenth of what the calculations indicated it ought to be. The mountains acted on the plumb bob almost as though they were not there. This could happen either if the mountains were composed of rock less dense than that under the lowland plain, which is improbable, or because the low-density surface rock extends to greater depth under the mountain than under the lowland plain. The latter is now known to be the correct interpretation.

This can be understood in terms of ice floating in water. Imagine two icebergs whose tops and bottoms are flat, but whose thicknesses are different. The height to which each stands above the water level depends upon the difference in density between ice, which is 0.9, and water, which is 1.0. So, as is well known, nine-tenths of the height of each iceberg is below water level and one-tenth is above water. The thicker iceberg, which is analogous to the mountains, stands higher than the thinner one, the lowland plain, not because its density is less but because it extends deeper—it has a "root." Mountains are now known to have roots. This knowledge comes from the study of elastic waves known as earthquakes, which have passed through the mountains. The crustal blocks with their roots "float" in the plastic material of the asthenosphere.

But the icebergs illustrate another point more important to the subject under consideration. If another layer of ice could be added to the top of either iceberg, it would sink a little, but its top would be a little higher, or, if a layer were removed from either, it would rise a little but its top would be a little lower. The exact relationship is that one-tenth the thickness of whatever amount is added or subtracted will be the rise or fall of the iceberg because at all times one-tenth of its total thickness is above the water level.

The same thing happens to the earth's crust. The equivalent of removing a layer of ice is the lowering of a high-standing area by erosion—this is "unloading." The equivalent of adding a thickness of ice, or "loading," is the deposition of sediment or volcanic material on a low-standing area. The change in "load" is "compensated for" by the plastic flow of the weak material of the asthenosphere. When fully "compensated"—that is to say, when balance is restored—the areas involved are said to be in isostatic adjustment. These relationships are illustrated and explained in fig. 71.

The boundaries of blocks that are in isostatic adjustment are rather vague, and the size of the smallest area that can become balanced is not known. Yet clearly the earth is extremely sensitive to loading and unloading. The response of the earth to the appearance during the Quaternary period of large masses of ice called continental glaciers caused depression of the crust amounting to hundreds of feet, with corresponding rises beyond the limits of ice.

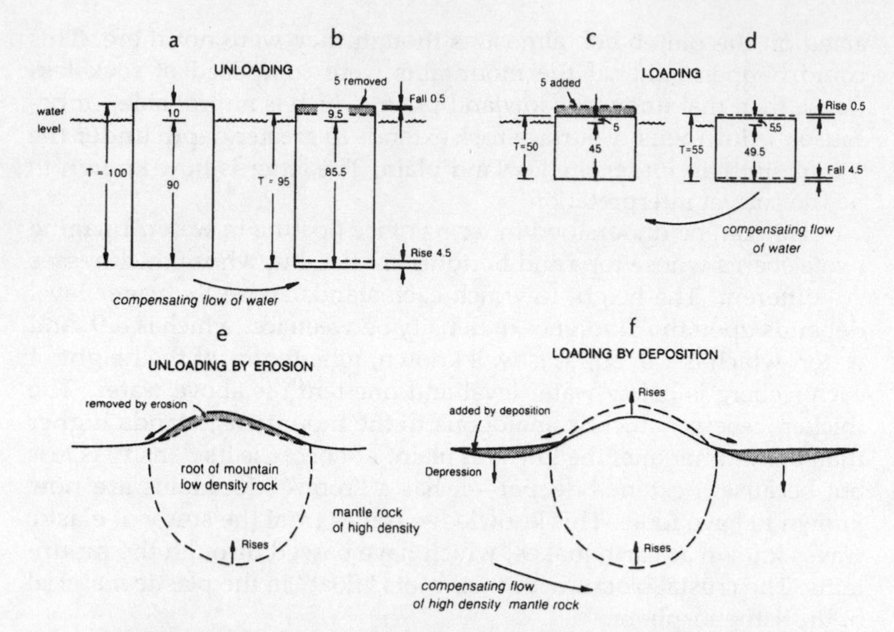

Fig. 71. An explanation of isostasy. *a,* The figure represents an ice sheet 100 feet thick floating in water. Ninety feet of it are below the water level and 10 feet of it are above, owing to the difference in density between ice, 0.9, and water, 1.0. *b,* The effect of removal of a thickness of 5 feet of ice from the top of *a.* The remaining thickness is 95 feet, of which 85.5 feet are below water and 9.5 feet are above. The top was lowered by 0.5 foot and the bottom raised by 4.5 feet. *c,* The initial condition of another ice sheet 50 feet thick, of which 45 feet are below water and 5 feet are above. *d,* The effect of adding to *c* the 5-foot thickness removed from *a.* The new thickness is 55 feet, of which 49.5 feet are below water and 5.5 feet are above. The bottom sank by 4.5 feet and the top rose by 0.5 foot. Note that the rise and fall of the two ice sheets, one unloaded and the other loaded, do not depend on the thickness of ice, but only on the change in thickness and the difference in density between ice and water. The rise and fall are accommodated (compensated) by the flow of water beneath the ice sheets. *e* and *f* show by analogy how a mountain with a root of rock of the same low density as that at the surface, embedded in mantle rock of higher density, will rise when eroded, and how adjacent areas will sink when they receive the sediment from the mountain. The changes in level are accomplished (compensated) by the plastic flow of mantle rock in the same way that the changes in level of the ice sheets are compensated by the flow of water.

It has also been shown that unloading of the crust by the evaporation of the waters of Lake Bonneville, the former very much larger lake of which Great Salt Lake is the remnant, resulted in the rise of the central part of the basin relative to the margins. And, in recent years, loading of the crust by the filling of Lake Mead behind Hoover Dam caused a measurable depression of the lake and surrounding country that is still continuing.

We can now consider what happened when the crust of the earth was loaded with the thick pile of rock of the Nevadan geosyncline. It is a simple matter to calculate what thickness of the mantle rock had to move for the load to be compensated if one knows the densities of the materials involved. Of course, some uncertainty is involved in selecting a value for the average density of the pile of rock, mostly of fragmented andesite, the pore spaces of which are filled with sea water. A fair guess at that value is 2.5, a value for many sandstones with about 20 percent of pore space filled with water. That which was displaced in the process of compensation is mantle rock with a density of 3.3, a value established through the study of the elastic waves of earthquakes. Using those values, 54,000 feet of the water-saturated andesite would displace 41,000 feet of mantle rock; that is, it would depress the initial sea floor by that amount. The difference of 13,000 feet is the amount by which the sea floor would be raised. In this case the new surface would be dry land 500 feet above sea level. Of course these calculations are only approximations, but they indicate that a sea of average depth can accept a thickness of sediment of about 53,000 feet, or 10 miles, before it is filled. The subsidence would be gradual as the rock accumulated, so one can understand that it is possible for a nearly 10,000-foot-high volcanic pile consisting of the Arlington, Goodhue, and Peale formations to be above, or on top of, a possible volcano nearly 11,000 feet high of the Taylor Formation.

Subsidence by isostatic adjustment cannot continue indefinitely. If sediment continues to accumulate, the sea will eventually be filled. The fact that truly shallow water deposits are present only in the upper part of the pile among the rocks of Triassic and Jurassic age in the Taylorsville District and in the Monte de Oro Formation near Oroville may be taken to indicate that the construction of the geosyncline was nearly completed.

Additional subsidence can be accounted for by the deformation that accompanies subduction. Subduction probably continued during the entire period of sedimentation and was, no doubt, responsible for the unconformity at Wades Lake, which shows that the Shoo Fly Formation was deformed, brought above sea level, eroded, and again submerged. Subduction was probably again responsible for a similar episode of a later date in the Taylorsville District, where rocks of Triassic and Jurassic age rest unconformably on rocks of Permian age, and, perhaps for other similar events that have not yet been discovered.

The rocks of the western belt are thought to have been brought to the geosyncline at the western edge of the North American Plate by the relatively eastward motion of the oceanic plate that lay to the west. The Feather River Serpentinite may mark the position of another old subduction zone that had been active during the early part of the Paleozoic era when the Sierra Buttes and the Elwell and Taylor formations were deposited.

That subduction temporarily prevails over isostatic adjustment is known from the observation that isostatic adjustment is imperfect or incomplete in many places along presently active subduction zones where volcanoes occur and earthquakes are prevalent. The combined processes of erosion and sedimentation, isostatic adjustment, and subduction to which obduction and volcanism are related are the processes that caused and permitted the accumulation of the great thickness of rock that is the geosyncline.

The thickness of 10 miles of the rocks of the eastern belt is clearly not a limiting value. Because so little is yet certain about the western rocks, the actual thickness could be as much as 20 miles. My best estimate is 13 miles, the value that developed in constructing the diagrams of fig. 72. The width of the area of deposition is not accurately known but very likely it was at the beginning almost 700 miles. The area of accumulation extended from southeast Alaska into Mexico, and although the rocks are not everywhere the same age, all are of the same general character. This mass of rock of almost staggering dimensions and volume comprised the Nevadan geosyncline.

SUMMARY

The condition that existed at the beginning of the construction of the Nevadan geosyncline was that of open ocean adjacent to the western shore of North America. The shore was in what is now western Utah. Not much can be said about the sea other than that it was open. It is shown in fig. 72, section A–A′, as being as deep as the average depth of the ocean today, about 2.5 miles. The width of the sea that would become involved in the geosyncline was about 700 miles, as is shown in fig. 72, section A–A′, to set the scale of future developments.

The organic sediments limestone and dolomite, and shale and sandstone derived from the land to the east, were deposited outward from the shore for a distance of perhaps more than 200 miles. Eventually their thickness became as much as 7 miles, as shown in fig. 72, section B–B′. All were deposited in shallow water because the sea floor subsided isostatically in response to the load of sediment.

Early in the Paleozoic era a subduction zone developed far out in the sea. Volcanic activity on the eastern plate above the subduction zone gave rise to the Devonian-Pennsylvanian arc system that consists of the Sierra Buttes, Elwell, Taylor, and Peale formations, whose total thickness is about 19,000 feet or nearly 4 miles. Volcanic sediments became widespread in the basin east of the arc and interfingered with the nonvolcanic and organic sediments that came from the east in the region that is now central Nevada.

Subduction ceased on that zone and began again on a new zone farther west. Volcanic activity again developed on the eastern plate above the subduction zone, which resulted in the deposition of the Arlington, Goodhue, and Reeve formations, as well as other rocks of Triassic age, as a new Permian-Triassic arc on top of the rocks of the Paleozoic arc system. Rocks were deposited as far east as central Nevada, and possibly as far west as what is now the Sacramento Valley. The thickness is at least 20,000 feet.

Again, that subduction zone ceased to operate and a third one developed still farther west, perhaps near the center of the present Sacramento Valley. Volcanic activity associated with this zone gave rise to the extensive rocks of Jurassic age that are about 15,000 feet thick at Taylorsville. Their total thickness was once greater because the top has been lost as a result of the intrusion of plutons of granitic rock and erosion. The Jurassic rocks extend well into Nevada. Late in the history of the Jurassic, the Smartville Complex was emplaced from the west by the still-obscure process of obduction.

Thus three volcanic arc systems are superimposed by deposition associated with three subduction zones that stepped in succession to the west. The combined thickness of the three arc systems is roughly 55,000 feet, or 10 miles. To that must be added 10,000 to 15,000 feet or more of the Shoo Fly Formation to give a total thickness of the geosyncline of 65,000 or 70,000 feet, about 12 or 13 miles. The total cannot be less and it may be more, for there is no way at present to account for the rocks of the central belt.

The conditions late in the Jurassic period are indicated in fig. 72, section B–B′. The geosyncline was completed after 250 million years of virtually continuous volcanic activity above the subduction zones. The accumulated material, perhaps 13 miles or more thick, was accommodated partly by isostatic adjustment and partly by deformation as the two crustal plates pressed upon each other.

The ocean shoreline had progressed westward to a position perhaps near the center of the Sacramento Valley. Much deformation had occurred during the deposition of the geosynclinal rocks, and there had been intrusions of magma as well as volcanic activity.

The Nevadan geosyncline, about 700 miles wide and 13 miles deep, was perhaps 3,500 miles long, extending from Alaska into Mexico. The stage had been set for the Nevadan Orogeny.

Fig. 72. Evolution of the Ancestral Sierra Nevada. *A–A′*, The initial condition in the Ordovician or Silurian period. The shoreline of North America was in western Utah. An open ocean of average depth of about 12,000 feet extends farther than 700 miles to the west. The curvature of the earth is not represented. The vertical scale is twice that of the horizontal scale. *B–B′*, The Nevadan geosyncline completed near the end of the Jurassic period. Nonvolcanic sedimentary rock derived from the east has accumulated on old continental crust to a thickness of about 7 miles. Volcanic rock and volcanic sediment about 13 miles thick has accumulated in the west. This was the result of melting along one or more (possibly three) subduction zones, to construct three island-arc volcanic ridges, each superimposed on the previous one. The width is 700 miles. The scale is the same as in *A–A′*. The shoreline has migrated to the site of the future Sacramento Valley. *C–C′*, The geosyncline deformed in the Nevadan Orogeny. The rocks have been folded, faulted, and metamorphosed. The width of the geosyncline is interpreted to have been reduced by about half, from 700 miles to 350 miles. The shortening was accomplished by folding and thrust faulting. The bottom of the geosyncline has been depressed to a depth of possibly 25 to 30 miles below sea level. The land surface is shown elevated to about 20,000 feet above sea level, but many thousands of feet of rock have already been removed by erosion. The western shoreline is now near the center of the future Sacramento Valley. The horizontal scale of *C–C′* and *D–D′* is twice that of *A–A′* and *B–B′*, and the vertical scale is double that of the horizontal scale. *D–D′*, The end of the Nevadan Orogeny. Silicic magma generated at depth intruded upward to form plutons of granitic rock. The intrusions, emplaced during the Jurassic and Cretaceous periods, are shown as not yet exposed at the surface by erosion. The metamorphosed nonvolcanic and volcanic rock and sediment now comprise new continental crust 25 to 30 miles thick added to the west side of the continent. The ocean shore has moved permanently to the west.

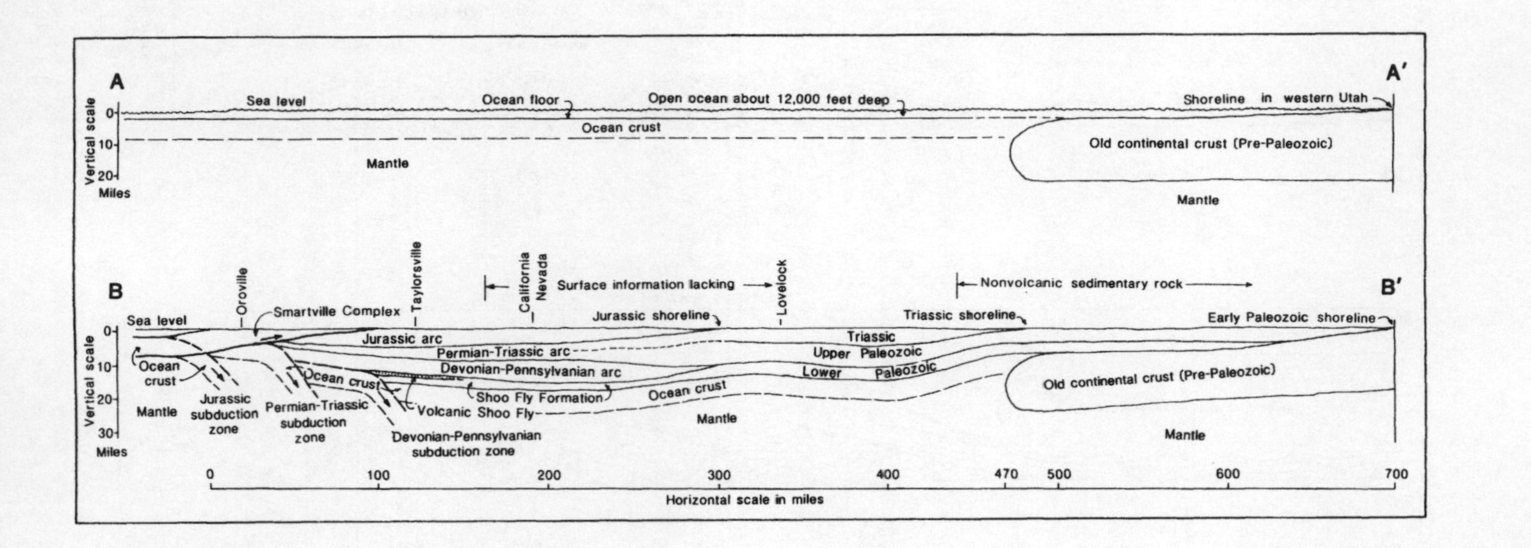

A
A′
Sea level
Ocean floor
Open ocean about 12,000 feet deep
Shoreline in western Utah
Ocean crust
Old continental crust (Pre-Paleozoic)
Mantle
Mantle
Vertical scale
0
10
20
Miles
B
B′
Oroville
Taylorsville
California
Nevada
Lovelock
Surface information lacking
Nonvolcanic sedimentary rock
Sea level
Smartville Complex
Jurassic shoreline
Triassic shoreline
Early Paleozoic shoreline
Jurassic arc
Permian-Triassic arc
Devonian-Pennsylvanian arc
Triassic
Upper Paleozoic
Lower Paleozoic
Ocean crust
Ocean crust
Ocean crust
Shoo Fly Formation
Volcanic Shoo Fly
Old continental crust (Pre-Paleozoic)
Mantle
Jurassic subduction zone
Permian-Triassic subduction zone
Devonian-Pennsylvanian subduction zone
Mantle
Mantle
Vertical scale
0
10
20
30
Miles
0
100
200
300
400
470
500
600
700
Horizontal scale in miles

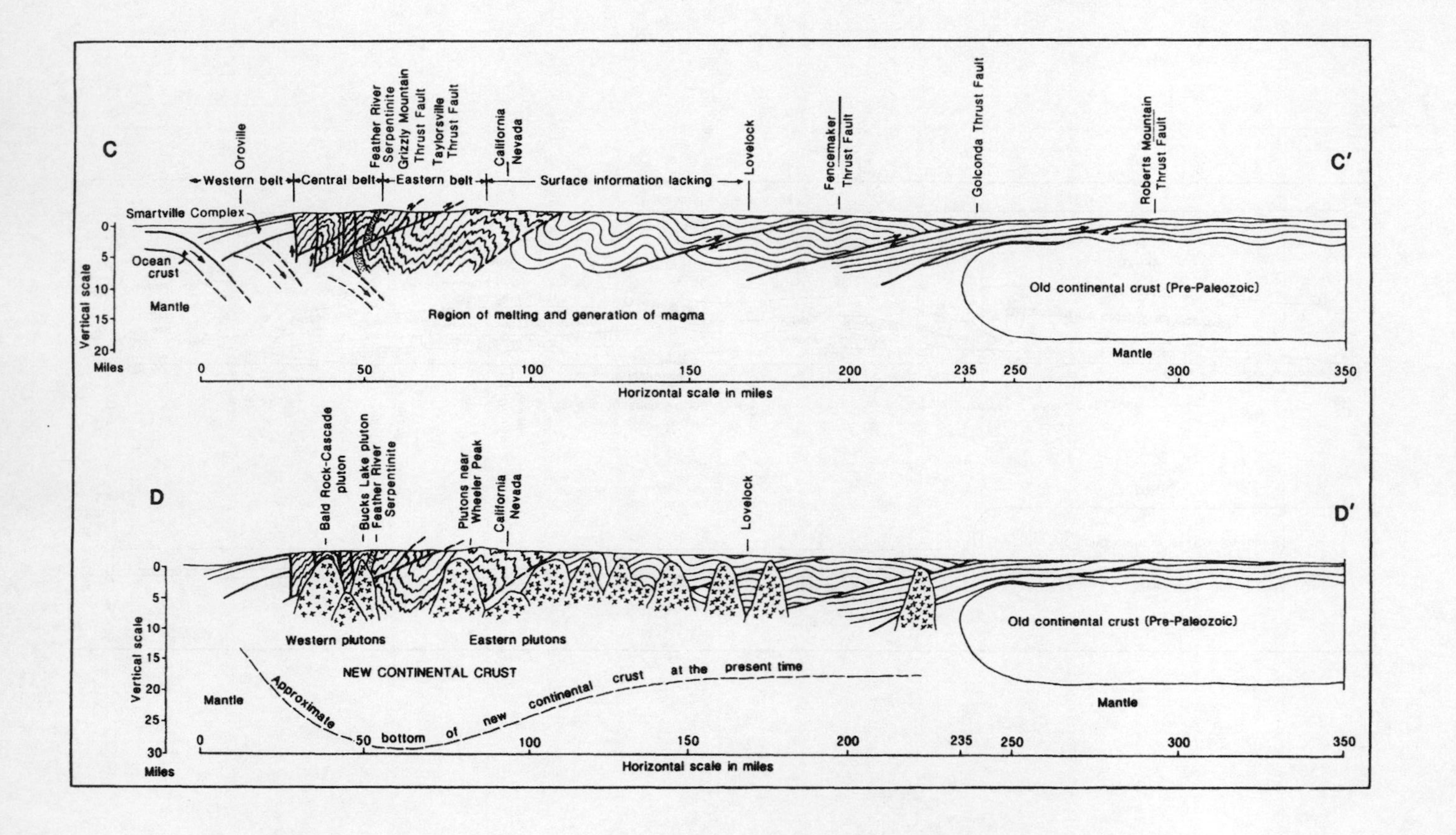
C
C′
Oroville
Feather River Serpentinite
Grizzly Mountain Thrust Fault
Taylorsville Thrust Fault
California
Nevada
Lovelock
Fencemaker Thrust Fault
Golconda Thrust Fault
Roberts Mountain Thrust Fault
Western belt
Central belt
Eastern belt
Surface information lacking
Smartville Complex
Ocean crust
Mantle
Region of melting and generation of magma
Old continental crust (Pre-Paleozoic)
Mantle
Vertical scale
0
5
10
15
20
Miles
0
50
100
150
200
235
250
300
350
Horizontal scale in miles
D
D′
Bald Rock-Cascade pluton
Bucks Lake pluton
Feather River Serpentinite
Plutons near Wheeler Peak
California
Nevada
Lovelock
Western plutons
Eastern plutons
NEW CONTINENTAL CRUST
Approximate bottom of new continental crust at the present time
Mantle
Old continental crust (Pre-Paleozoic)
Mantle
Vertical scale
0
5
10
15
20
25
30
Miles
0
50
100
150
200
235
250
300
350
Horizontal scale in miles

4

The Nevadan Orogeny and the Ancestral Sierra Nevada

Orogeny is the name applied to an episode of deformation of a part of the earth's crust that results in what are known as folded mountains. It is the ultimate consequence of a collision between major crustal plates. More particularly, it is an inclusive term for all the processes that result in the development of the typical features of folded mountains: the folding and faulting of the layered rocks of a geosyncline, their metamorphism, the production of magma, and the intrusion of plutons. Finally an elevated terrain that is called mountainous, such as that of the Alps of Europe, is produced. On the other hand, the term must be applied with care, for although the Appalachians, for example, are a folded mountain system, their present elevation is not a result of the orogeny, but of a much younger event of a different kind.

All elevated or mountainous terrains have been or will be eroded away, but the folded and faulted metamorphosed rocks and the plutons that are the products of orogeny remain in the earth's crust indefinitely long after the mountainous land has disappeared. Their presence is evidence that an orogeny has occurred. Thus the high-standing land mass that we call the Sierra Nevada is not a folded mountain system and it is not the result of orogeny. It is, however, a small part of an older folded mountain system, often called the Ancestral Sierra Nevada, that was of far greater extent than is the present range. The Ancestral Sierra Nevada was the product of the Nevadan Orogeny and whatever mountains that resulted from it had been eroded to an area of low relief long before the present range was elevated in a quite different manner. That story is told in later chapters.

In Chapter 3, I related the story of the construction of the Nevadan geosyncline, an enormous mass of sediment, mostly volcanic, originally about almost 700 miles wide, perhaps 3,000 miles long, 12 to 13 miles thick, and possibly as thick as 20 miles. Its cross-sectional form in late Jurassic time is indicated in fig. 72, section B–B′.

The rocks of the geosyncline had undergone some deformation during accumulation as evidenced by the unconformity between the Shoo Fly and Sierra Buttes formations at Wades Lake, by the unconformity between the Triassic and Jurassic rocks in the Taylorsville District, and by emplacement of the Smartville Complex.

Now, however, late in the Jurassic period a paroxysmal deformational event occurred. Deformed as though caught in the jaws of a giant vise, the rocks were folded, faulted, metamorphosed, and intruded by magmas that became plutons, all in a geologically short period of time. The jaws of the vise were the North American Plate, with the North American continent on the east, and the oceanic plate on the west. The two moved against each other with much greater violence than at any previous time. The consequent deformation resulted in folds and faults on all scales from the microscopic to the gigantic.

Interpretations of the large structures that resulted from the deformation are illustrated in figs. 72, sections C–C′ and D–D′, 73, and 74. The first two are broad generalizations, but the structure shown in fig. 73 is realistic. It is the result, slightly modified by me, of the careful study of the region around Taylorsville by V. E. McMath. Were another geologist to repeat the work done by McMath his results would probably not differ in any essential way. Such large structures cannot be viewed in the field. But that they are constructed on paper from the data collected in the field and recorded as a geologic map makes their existence no less real.

In figs. 41 and 73 one can see that the Grizzly Mountain Thrust Fault separates two belts of rock, in each of which the sequence of Paleozoic rocks from the Shoo Fly Formation to the Reeve is present. In each sequence the rocks face toward the east although at the ground level they are inclined to the west. That is to say that they are overturned or that the originally nearly horizontal layers passed through the vertical when transfered to their present position. That such an

Fig. 73. A geologic cross section of a part of the Ancestral Sierra Nevada from near Keddie to near Wheeler Peak. Modified from the unpublished work of V. E. McMath. The line of section is shown on fig. 41 between points B and C. This is a realistic, conservatively constructed cross section based on careful field studies.

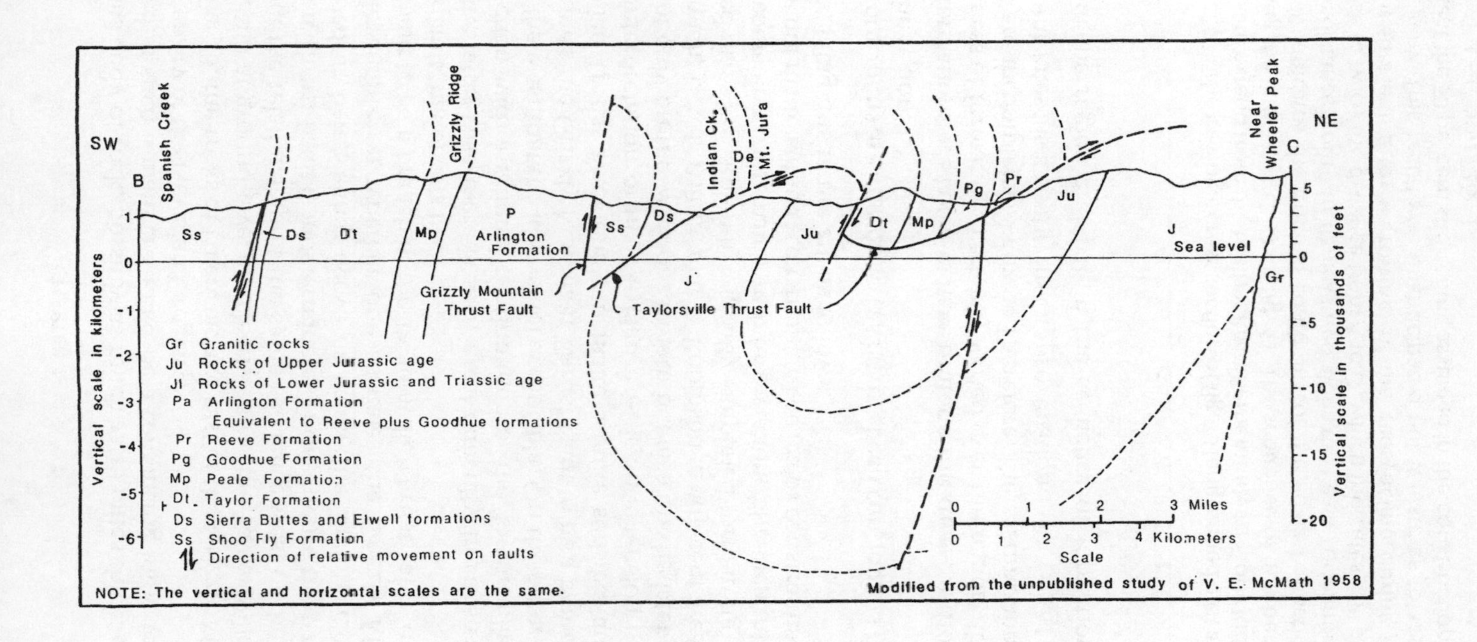
SW
B
Spanish Creek
Grizzly Ridge
Indian Ck.
De
Mt. Jura
Near
Wheeler Peak
C
NE
Ss
Ds
Dt
Mp
P
Arlington
Formation
Ss
Ds
J
Ju
Dt
Mp
Pg
Pr
Ju
J
Sea level
Gr
Grizzly Mountain
Thrust Fault
Taylorsville Thrust Fault
Vertical scale in kilometers
1
0
-1
-2
-3
-4
-5
-6
Vertical scale in thousands of feet
5
0
-5
-10
-15
-20
Gr Granitic rocks
Ju Rocks of Upper Jurassic age
Jl Rocks of Lower Jurassic and Triassic age
Pa Arlington Formation
Equivalent to Reeve plus Goodhue formations
Pr Reeve Formation
Pg Goodhue Formation
Mp Peale Formation
Dt Taylor Formation
Ds Sierra Buttes and Elwell formations
Ss Shoo Fly Formation
Direction of relative movement on faults
0 1 2 3 Miles
0 1 2 3 4 Kilometers
Scale
NOTE: The vertical and horizontal scales are the same.
Modified from the unpublished study of V. E. McMath 1958

enormous body of rock, possibly as much as 13 miles thick, could be deformed in such a manner should be a convincing indicator of the great intensity of the Nevadan Orogeny.

In fig. 73, the rocks of Paleozoic age east of the Grizzly Mountain Thrust Fault at ground level are clearly above the Taylorsville Thrust Fault, but the relationship of the same sequence west of the Grizzly Mountain Thrust Fault is indeterminate. The sequence there could be interpreted as being either above or below the Taylorsville Thrust Fault, depending on whether one assumes the Grizzly Mountain Thrust Fault to be a branch of the Taylorsville Thrust Fault, or to be a younger fault that offsets it.

In the latter case there is such great difficulty in constructing a cross section with a rational continuation of the Taylorsville Thrust Fault to the west that the idea is not acceptable. On the other hand, if one assumes the Grizzly Mountain Thrust Fault to be a branch of the Taylorsville Thrust Fault one can make a quite sensible and fairly simple interpretation of fig. 74. Figure 74 is an extension of fig. 73 based not only on the work of McMath but also on the later work of J. L. Hannah, J. A. D'Allura, and L. Robinson in adjacent areas. Their combined maps are shown in a greatly simplified form in fig. 41.

Of course fig. 74 remains an interpretation, but such interpretations are not made without constraints. The major constraints applied in constructing this figure are as follows:

1. The formations are interpreted as having been continuous throughout the region.
2. The formations are taken to have uniform thickness. Although the thickness of formations certainly varied over the region, the variations are incompletely known. Therefore, the assumption of uniform thickness is a useful first approximation.
3. Within the region and within the sequence of formations from

---

Fig. 74. The geologic cross section of fig. 73 lengthened east and west to extend from the Feather River Serpentinite to Stony Ridge along the line A–B–C–D on fig. 41. The sequence of numbers 1 to 10 indicates the original continuity of the Peale Formation. The distance from 1 to 10 measured along the Peale Formation is close to twice the distance from A to D, which indicates a 50 percent shortening (or narrowing) of the geosyncline during the orogeny. This result was unanticipated but it justifies to some extent the 50 percent shortening assumed in the construction of figs. 72*C–C′* and 72*D–D′*.

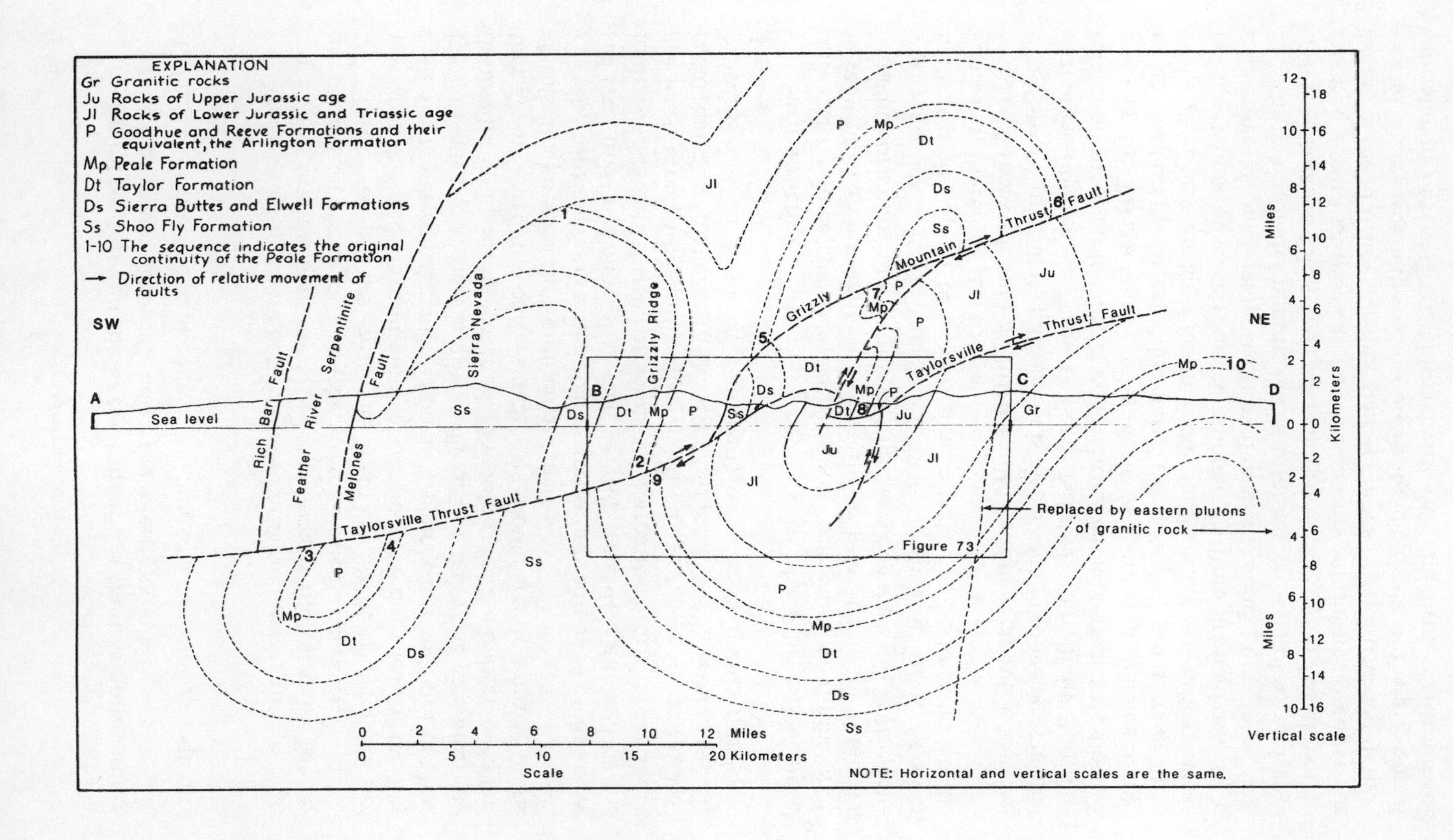
EXPLANATION
Gr Granitic rocks
Ju Rocks of Upper Jurassic age
Jl Rocks of Lower Jurassic and Triassic age
P Goodhue and Reeve Formations and their equivalent, the Arlington Formation
Mp Peale Formation
Dt Taylor Formation
Ds Sierra Buttes and Elwell Formations
Ss Shoo Fly Formation
1-10 The sequence indicates the original continuity of the Peale Formation
→ Direction of relative movement of faults
SW
NE
A
B
C
D
Sea level
Rich Bar Fault
Feather River Serpentinite
Melones Fault
Sierra Nevada
Grizzly Ridge
Grizzly Mountain Thrust Fault
Taylorsville Thrust Fault
Replaced by eastern plutons of granitic rock
Figure 73
Scale
Miles
Kilometers
Vertical scale
NOTE: Horizontal and vertical scales are the same.

Sierra Buttes through the Reeve formations there is no unconformity that would materially affect the interpretation.

4. Each anticline was flanked by a syncline in a continuous sequence of folds.

These constraints are both imposed and justified by the geologic maps. The original continuity of formations is indicated by the sequence of numbers 1 to 10 attached to the Peale Formation.

The extension of the great syncline to a depth of 8 or 9 miles is not unreasonable and is really minimal because the thickness of the formations may be greater than shown. Certainly the base of the Shoo Fly Formation is at least another mile or two below its top, and, undoubtedly, complications of the fold would carry the bottom even farther down. The depth of the syncline as shown is only a fourth of the 40-mile thickness of the continental crust.

The construction carried to a height of more than 14 miles above sea level does not mean that such heights were ever attained by the folded and faulted rock for as soon as deformation brought rock above sea level, erosion began its attack. Although erosion is usually less rapid than deformation, otherwise no mountains could ever exist, it is probable that the resulting mountains were never higher than 3 to 5 miles.

Such structures as those of fig. 74 could have been developed only during some considerable period of time, and, no doubt, they developed in an intermittent or episodic manner. Certainly there is a sequence of faults. One in the syncline is older than the Taylorsville Thrust Fault because it seems to terminate upward against it. Another fault near the center of the syncline offsets the Taylorsville Thrust Fault and, therefore, it is the younger.

The Grizzly Mountain Thrust Fault is younger than the Taylorsville Thrust Fault but its movement is also a second movement in that part of the Taylorsville Thrust Fault west of their point of division. The total movement on the Taylorsville Thrust Fault (between numbers 3 and 2 on fig. 74) is about 12 miles. Perhaps as much as 6 miles of that is the offset on the Grizzly Mountain Thrust Fault (from number 7 to 6 in fig. 74). These are large faults. Other faults range downward in size to those whose offset can be accommodated within the field of view of a microscope. Small faults are shown in figs. 12 and 14.

The folds of the region, like the faults, are complex, and successively smaller folds are superimposed on larger folds down to those so small that they can be studied only with a microscope. For example, the anticline-syncline pair at Wades Lake, illustrated in figs. 49 and 50, is superimposed on the east-facing side of the great anticline shown west of Grizzly Ridge in fig. 74. (The former continuity of the formations from there south to Wades Lake is indicated in fig. 41.) Many other folds in the same position and of the same amplitude, 200 to 300 feet, can be examined between Mt. Elwell and Sierra Buttes, affecting the Sierra Buttes and Elwell formations. Also at Wades Lake where one can clamber over bare rock to examine the folds in detail, folds with an amplitude of only a few feet are superimposed on the crest of the anticline, as shown in fig. 49. Folds of that size and smaller are common throughout the Lakes Basin District, especially in the Elwell and Peale formations and particularly in chert.

Folds of the kind called isoclinal (meaning inclined in one direction), in which the two limbs are nearly parallel like the folds of a closed bellows, are common in many places (see figs. 11 and 44) but they are inconspicuous and difficult to recognize. Like open folds they also range in size from microscopic to large.

The descriptions of folds and faults set forth above, together with the photographs that illustrate Chapters 1 and 3, exemplify the results of the Nevadan Orogeny. Although data are still insufficient to permit expansion of fig. 74 to encompass the entire width of the range, one may expect that the structural character elsewhere is not entirely dissimilar. Farther west, large anticlinal and synclinal folds and large steeply inclined faults are present, as shown on figs. 40 and 72, section C–C′. Gently inclined thrust faults similar to the Taylorsville Thrust Fault are surely present in the central belt. Steeply inclined faults may offset the thrust faults or branch from them as the Grizzly Mountain Thrust Fault branches from the Taylorsville Thrust Fault, or both relationships may occur.

Thus there is much similarity in the nature of folds and faults among the rocks of the three belts. However, the central belt contains rock called melange that is absent from the rocks of the eastern belt younger than the Shoo Fly Formation.

Melange is literally a mixture of rock that lacks the continuity of layers characteristic of the eastern belt rocks, where layers only a foot thick can be traced for more than a mile and formations are continuous for tens of miles. Melange is neither a varietal rock nor a formation. It is a complex, a jumble of once-layered rock disrupted by intense deformation so that the layering is destroyed. Weak portions of the originally layered sequence become a matrix of small to minute pieces in which are embedded angular blocks of dimensions measured in feet to tens of feet that are fragments of originally thicker, harder, or stronger layers. The margins of melange are often faults, and indeed, melange may be thought of as akin to a large volume of fault-brecciated rock. Melanges of the central belt were once alternations of chert, sandstone, shale, volcanic ash, and rare limestone.

The Feather River Serpentinite is a slab of old rock faulted on both sides that lies between the central and eastern belts. The presence along its western edge of a strip of rock metamorphosed at higher temperature than is usual elsewhere may be taken to indicate that that was its lower side before being brought into its present position. Since it is now inclined to the west it appears to have been overturned, as are the rocks of the eastern belt adjacent to it (fig. 74). Since it is not typically ophiolite rock but is ultramafic like the material of the mantle, it was probably brought up from a depth of 15 to 20 miles.

Such an origin of the Feather River Serpentinite can be accounted for by imagining that it is the remainder or a fragment of the east limb of a very large overturned and faulted anticline, the axis of which is obscured by large-scale faulting on gently inclined thrust faults and by the pervasive deformation of the rocks of the central belt. Its position may be analogous to that of the Peale Formation, as reconstructed between numbers 7 and 8 on fig. 74, but on a far larger scale.

In the western belt the Smartville Complex, being of relatively strong rock, did not respond to deformation in the same way as did the rocks of the central belt. In fact, much of it shows scarcely any of the cleavage that pervades and characterizes the metamorphic rock of the Sierra Nevada elsewhere. Available evidence indicates that the surface below the Smartville Complex is a fault gently inclined to the west. What is beneath is unknown. However, Smartville rocks north of Oroville may rest on central-belt rocks separated by a thrust fault, as I have indicated on fig. 40. This is still a speculative interpretation.

In any event, being composed of ocean crust, the Smartville slab must have moved eastward on a thrust-fault surface, but how far east it may have extended is not clear. Presently, excepting possibly the area north of Oroville, as noted above, it is separated from the central belt by the steeply inclined Wolf Creek Fault. On the other hand, the fact that some rock of the central belt is similar to the Smartville and apparently of the same Jurassic age indicates that there may have been a former eastward extension of the Smartville that became involved in the intense deformation that affected the central belt.

It is quite apparent that compression reduced the width of the geosyncline represented in figs. 73 and 74. Folding and faulting were the mechanisms by which the shortening was accomplished. At the same time rock was depressed to greater depth and rock was also raised above sea level as the Ancestral Sierra Nevada.

The amount of shortening can be approximated from fig. 74. The distance along the folded rocks that were once essentially flat, measured on the base of the Peale Formation, is 85 miles; the equivalent horizontal distance is 45 miles. To that indicated shortening of 40 miles by folding must be added another 12 miles of motion on the Taylorsville Thrust Fault, so the shortening was more than 50 miles, considerably more than half the original span of 85 miles. No doubt additional amounts of shortening occurred on many still-undiscovered faults. Possibly the same proportion of shortening occurred in the entire 100-mile span of highly deformed rock from somewhere beneath the Sacramento Valley to the edge of the eastern plutons, as indicated in fig. 40, a distance that was once 200 miles. Perhaps such a proportionate shortening affected the entire width of the geosyncline.

The deformational history of the geosyncline farther east is in part problematic because across the next 100 miles, or as far as Lovelock, Nevada, the metamorphic rocks are partly concealed beneath a blanket of very much younger rock and have partly disappeared by replacement by plutons, as shown in fig. 72, section D–D′. East of Lovelock, the old rocks are at the surface again. There they are only slightly metamorphosed and the folds are open, so not much shortening can be attributed to folding. However, many large thrust faults gently inclined to the west are present in the next 120 miles. The movement on several is with good reason believed to amount to several tens of miles, and the total shortening may be as much as 100 miles. Considering that more such faults are probably present in the 100 miles between Plumas County and Lovelock, it seems not at all unreasonable that the shortening of the entire width of the geosyncline during the Nevada Orogeny was also by one-half. That is, therefore, the value I used in constructing fig. 72, section C–C′, to illustrate the effect of the Nevadan Orogeny on the entire width of the geosyncline.

Figure 72, section C–C′, is not a realistic cross section as are figs. 73 and 74, mostly for practical reasons. One could do better but the drawing would need to be greatly extended and at a much larger scale. I have not shown in it the same rock units used in fig. 72, section B–B′, because sufficient information is lacking. Instead I have used "form lines" that are not the boundaries of formations but are merely indicators of the direction of layering of the rocks. They do permit the representation of folds and the positions of faults.

In each part of fig. 72 the vertical scale is twice that of the horizontal scale. That also is a matter of practicality. To indicate the reduction of width of the geosyncline the horizontal scale of fig. 72, section C–C′, is half that of fig. 72, section B–B′. For example, the distance from Oroville to the edge of the old continental crust is 470 miles on fig. 72, section B–B′, and 235 miles on fig. 72, section C–C′. Both values are marked on the horizontal scale bars.

The descriptions of the effects of deformation in the preceding paragraphs relate mostly to large features of the rocks, those that can be seen in the field or that are revealed by making geologic maps. I have mentioned, however, that deformation plays a vital role in metamorphism. It affects rocks in detailed ways that are revealed by the use of the optical microscope, x-ray diffraction apparatus, and other instruments for the physical and chemical analysis of rocks and minerals.

The effect of deformation in the metamorphism is shown by the fact that, in places, the once angular blocks in volcanic breccia have been squeezed into the shape of lenses (see fig. 32). The presence of the cleavage in the metamorphosed rock is dependent on deformation. For example, it is possible to see at some places where cleavage crosses the boundary between two kinds of rock, that the boundary is offset by a minute amount across the cleavage surfaces; the cleavage surfaces are minute faults. The sum of many such small movements can result in what are called shear folds, as shown in fig. 33, in which a dike appears to be folded but has not been bent.

Grains of brittle minerals like quartz become crushed and their parts strewn out on cleavage surfaces and in many places it can be shown that grains were rotated while the cleavage developed. Heat generated by deformation was a principal source of the energy consumed in recrystallization during metamorphism although heat came from other sources as well. The temperature of the geosynclinal rocks began to rise immediately after deposition because they sank by isostatic adjustment to deeper levels in the earth where the temperature is normally higher. The heat acquired by burial is that which normally flows from the interior of the earth to the surface.[1]

Heat was also produced in the rocks of the geosyncline by the radioactive decay of atoms of uranium, thorium, and potassium that are normal components of the mineral grains of the rocks. And heat was transported from greater depth by rising bodies of magma that intruded the deformed geosyncline, giving rise to the plutons described in Chapter 5, and shown on fig. 72, section D–D′.

Compression of the geosyncline not only reduced its width by half, but it also depressed the geosynclinal rock far beyond its original depth of 12 or 13 miles. Affected by metamorphism and partial melting to magma that intruded to form plutons of granitic rock, the originally largely fragmental mass became hardened and strengthened into new continental crust. The depth to which the new continental crust extends is known from the study of the elastic waves of earthquakes and from the study of variations of the earth's force of gravity to be as much as 25 to 30 miles beneath the northern Sierra Nevada and Diamond Mountains, as shown on fig. 72, section D–D′. The structure at that depth is not known but is probably like that at lesser

1. The temperature in sedimentary rocks penetrated by deep wells in search of petroleum increases at the rate of about 1° F per 100 feet of depth.

depths, as shown on fig. 72, sections C–C′ and D–D′. This deeply depressed rock is the root of the Sierra Nevada and the mountain root of figs. 2 and 71.

The orogeny also elevated geosynclinal rock above sea level so that it immediately became attacked by weathering and erosion. The elevated portion became a range of folded mountains, the Ancestral Sierra Nevada, that, by analogy to the Alps or the Andes, may have been a lofty and rugged mountain range although virtually nothing is known about the landscape at that time.

The Pacific shoreline of North America had been progressively removed from the Nevada-Utah region to a new position far to the west. After a few million years it was again at the east side of the Sacramento Valley, but never since has it been east of the present Sierra Nevada.

The deformational part of the orogeny in northern California can be dated fairly closely. The youngest deformed fossiliferous rocks are of late Jurassic age. Those at Taylorsville are estimated to be about 150 million years and those in the western Sierra Nevada about 145 million years old. These dates are probably close to the end of deposition of the layered rocks. Deformation soon began but the exact time is not known. The layered rocks of the Nevadan geosyncline continue northward beneath young volcanic and sedimentary rock layers, and reappear in the Klamath Mountains, where they are of the same kind, show the same degree of metamorphism, and are intruded by similar plutons. There the Shasta Bally Pluton, a few miles west of Redding, has a radiometric age of about 130 million years, nearly the same as that of the Bucks Lake Pluton near Belden. The Shasta Bally Pluton had its cover removed by erosion, was again depressed below sea level, and had fossiliferous marine sedimentary rocks of Cretaceous age estimated to be about 118 million years old deposited on it. One concludes from this that the rocks of the geosyncline were deformed, metamorphosed, invaded by plutons, and sufficiently eroded to expose the granitic rock in a time span of probably not more than 27 million years, or less than 12 million years.

Unmetamorphosed fossiliferous rocks of late Cretaceous age, probably not more than 80 million years old, rest on metamorphic or granitic rocks of the Sierra Nevada along Chico Creek, at Pentz north of Oroville, and at Folsom, but the data do not permit as close dating of the orogeny as do the rocks west of Redding. So, we conclude that the orogeny began about 140 million years ago and that deformation was completed by about 118 million years ago, or, perhaps, it lasted about 22 million years.

However, as noted in Chapter 5, the period of invasion of the plutons lasted considerably longer—about 55 million years.

5

The Plutons: Their Nature, Origin, and Age

The development of magma and the intrusion of plutons of granitic rock are part of the process of orogeny. Plutons comprise more than half the entire area of the Sierra Nevada and most of that part south of the Tuolumne River. While less extensive at the present erosion surface of the northern part of the range, plutons are probably essentially continuous at depth, as indicated in fig. 72, section D–D′. Thus they constitute a second great group of rocks of the Sierra Nevada in area, in volume, and in importance. Their relationship to the metamorphic rock is also shown in fig. 72, section D–D′.

The term granitic is used here as a general varietal name for a family of rocks that are plutonic and that have many features in common with the rock granite, which is, of course, one of the family. Granite as strictly defined is not common in the Sierra Nevada. The other rocks of the group differ from granite in varying degree in their mineralogical and chemical composition but all of them are coarse grained—that is, the individual grains are easily visible with the unaided eye—and most of them have rather uniform grain size. Most are speckled light and dark, as in the example in fig. 17; in most of them light-colored grains are much more abundant than dark-colored, but some have more dark- than light-colored grains (see fig. 18). Most of the granitic rocks of the Sierra Nevada are light colored, white to gray, and contain in abundance the glassy-appearing mineral quartz. The shiny black mica biotite is the most conspicuous dark mineral, but the less lustrous black hornblende is also very common.

Some plutons are roughly circular in outline (figs. 40 and 75), but others are elongate and some are irregular in shape. Many are composed of more than one variety of granitic rock; in many instances the

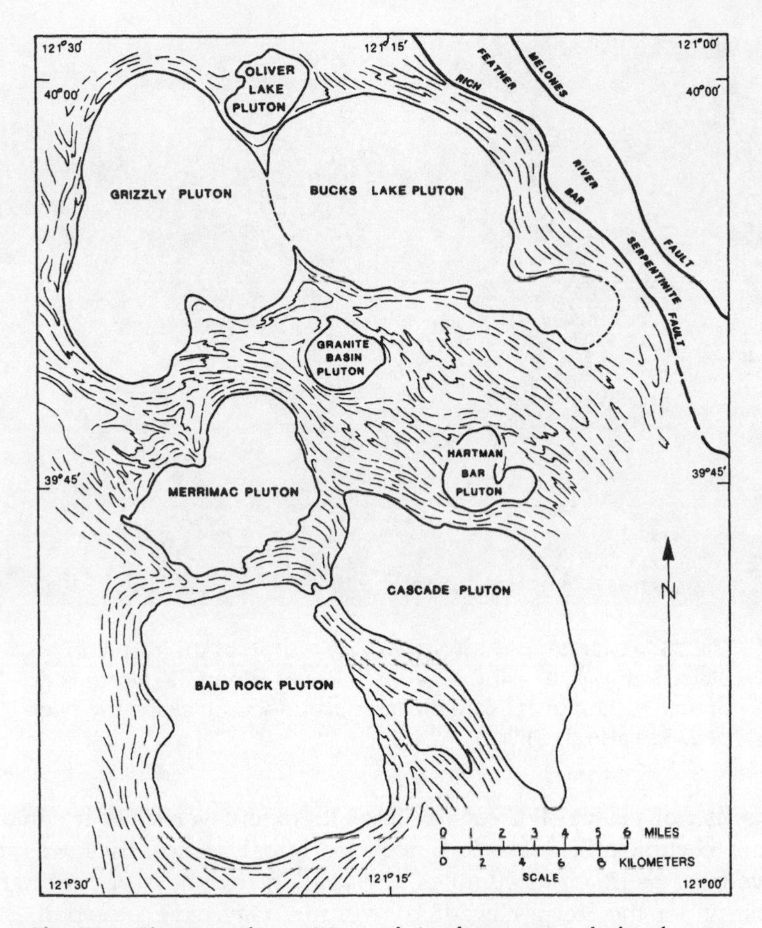

Fig. 75. Plutons of granitic rock in the western belt of metamorphic rocks, and a generalized representation of the apparent structure of the metamorphic rocks. The latter appear to have been thrust aside by the intruding magma. After the published studies of R. R. Compton (1955) and A. Hietanen (1951, 1973). Compare this figure to fig. 40.

rock of the central part contains more silica and potassium than does that of the outer part. At the margin of many plutons there is a narrow rim that is notably richer in dark minerals than is the interior. Dikes of granitic rock that are extensions of the pluton are usually present in the metamorphic rocks adjacent to the plutons.

Several of the plutons in the western part of the Feather River basin, shown in figs. 40 and 75, have been studied in some detail. One of these, the Grizzly Pluton (fig. 76) has an area of about 55 square miles. Its top has been eroded away and its bottom, as usual with plu-

Fig. 76. Granitic rock of the Grizzly Pluton at the rest stop on State Highway 70 near Cresta (Arch Rock), along the North Fork of the Feather River. Note the large, rounded blocks in the river bed. See also fig. 17.

tons, is not visible. For each mile of its extent vertically it contains some 55 cubic miles of rock. Since its original vertical thickness may have been as much as 40 miles, the known thickness of continental crust under the Sierra Nevada, its volume may have exceeded 2,000 cubic miles. That may seem to be a very large volume, but some plutons are even larger.

The area of granitic igneous rocks in the eastern part of Plumas County, shown in fig. 40, has not been studied in detail. It is part of a vastly larger area that is undoubtedly a complex of plutons whose size, shape, and character remain to be determined. This larger area continues north of Susanville beneath younger rocks; on the south it extends to Lake Tahoe and beyond, and on the east it extends well out into the state of Nevada. Much of it is covered by younger rocks.

An excellent place to observe the rocks of a pluton is along State Highway 70 in the Feather River Canyon between Pulga and Rock Creek Dam, where for a distance of 16 miles the river and highway cross the Bucks Lake and Grizzly plutons (figs. 40 and 76). In many places the rocks are not very weathered and can be examined in detail. The rocks of the eastern area are mostly weathered, but fresh rock can be seen on State Highway 70 between Blairsden and Portola a little west of the bridge across Willow Creek. Deeply weathered granitic rock is present on State Highway 70 east of Portola, and on State Highway 89 from Calpine to the summit of the grade toward Graeagle (see fig. 5). Weathered granitic rock is also present at Yuba Pass on State Highway 49 between Sattley and Sierra City.

It is now a generally accepted hypothesis that the magma that forms the plutons originated along a subduction zone. The magma must then migrate upward into the rocks of the deformed or deforming geosyncline. The magma could result from the partial melting of the oceanic crust[1] that is plunging into the mantle and becoming hotter as it goes deeper or from the partial melting of the geosynclinal rocks; or it could be a mixture of both; or some or all of it could come from the mantle. Some plutons in the central Sierra Nevada show evidence of having been deformed; therefore, they are old enough to have been involved in the deformational part of the orogeny. However, most are not deformed and the invasion of plutons continued until well after the deformation was complete.

A major problem in geology concerns the processes by which magma rises and the manner in which space is provided for a pluton. Four concepts of how the space is made have prevailed during this century: (1) the magma is forced or injected upward and the rock that formerly occupied the space is pushed aside; (2) the magma melts, or chemically digests, the rock; (3) the magma mechanically removes the rock by dislodgement of it, piece by piece, by intruding along cracks in the rock; (4) there was no magma—the pluton is the result of extreme metamorphism by which the original rock is transformed into plutonic, igneous-appearing rock without melting. The last theory is indefensible and is not considered any further.

One must keep in mind that the magma rises 20 to 40 miles from its place of origin, and that a considerable amount of time passes while that happens. It is reasonable to think that the magma and its wall rocks at different times and at different levels may react with each other in different ways depending on the local conditions. Also important is the fact that the level at which one examines a pluton depends on the amount of erosion since it finally crystallized, so that in viewing different plutons one may be viewing them at different levels or as if they were in different environments.

Even a casual glance at fig. 75 will impress the reader with the feeling that the plutons were injected; that the magma pushed aside

1. Silicate rocks do not melt at constant temperature to a liquid of the same composition, as does ice. They melt over a range of temperature to yield at first small amounts of liquid richer in silica and potassium than the solid rock from which it melts. In this way silica-rich magma can be derived from rocks much less rich in silica.

and deformed the adjacent layered metamorphosed rocks. The idea is even more attractive in the light of the fact that the cleavage in the metamorphic rock is also, for the most part, concentric to the margins of the plutons. There are irregularities to be sure, but it appears that the cleavage was present before the magma of the pluton arrived at the present level of exposure and that it also was pushed aside, although with some penetration of magma across it. The case has not been proved for any of these plutons although R. R. Compton expressed the opinion that three-fourths of the space occupied by the Bald Rock Pluton, which he studied in detail, was made by this process. Convincing evidence for emplacement by injection has been presented for plutons elsewhere in the world.

The second process, chemical digestion of the wall rocks, certainly has occurred at many plutons, and good examples are present in this region. Along the western limit of the large eastern area of plutonic rocks, especially between the Middle Fork of the Feather River and State Highway 70, and also high on Grizzly Ridge between Bull Run and Little Long Valley Creek northeast of Cromberg, there is a zone, in places as much as a quarter mile wide, of dark-colored rock that is transitional between the typical granitic rock and the metamorphic rocks that have cleavage. The transitional rock is rich in hornblende, its grain size is highly variable even within the limits of small samples, and bits of metamorphic rock can be recognized even though the boundaries are indistinct. It is obvious that it is the result of reactions between the silica-rich magma of the pluton and the less-silica-rich metamorphosed volcanic rocks of the wall. There are limits to the extent to which such chemical reactions can occur, depending on the compositions of the wall rock and the magma and on the availability of heat. It also seems necessary that the wall rock must have become fragmented, as along fractures such as joints, so that the process might involve large volumes of rock. For those reasons it is doubtful that the total space for plutons is made by chemical reactions alone.

The third process involves the removal of the wall rock in pieces by injection of magma along cracks or joints. Once liberated from the wall, pieces of rock, if sufficiently heavy, would sink through the magma under the influence of gravity. The process is called stoping, by analogy with the mining process of opening large underground rooms called stopes. Examination of the marginal parts of some plutons shows the reality of the process, for places are known where fragments of the metamorphic rocks of the wall are abundant, and blocks can be seen in positions that indicate that they were in the process of being dislodged when the act was suspended by crystallization of the magma. Some plutons, however, have few or no blocks of wall rock in them. There is, of course, the problem of the disposal of the blocks. If blocks removed in this way at all margins of the pluton simply sank into the magma there would be no net gain in space. So it may be that the process takes place when the plutons are near the upward limit of intrusion. On the other hand, blocks removed from the walls are in a favorable position for chemical digestion, and, as already pointed out, there is real evidence that such does occur. So there are three processes that can combine in various ways to provide room for plutons. They may work together, alone, or in combination, to produce different effects, at different times and different levels during the intrusion process. Compton estimated that one-fourth of the space for the Bald Rock Pluton was made by stoping and digestion of the stoped material.

Evidence of the process of reaction with fragments of the metamorphic rock can be seen near the tunnel on State Highway 70 at Grizzly Creek. There the granitic rock is near the margin of the pluton. It is dark colored, being rich in hornblende owing to reaction with the wall rocks. In the east side of the south end of the tunnel many fragments of metamorphic rock rich in red-brown garnet and other minerals are imbedded in the plutonic rock. The fragments were probably originally pieces of limestone like that which occurs nearby, but up the slope well above the river.

The processes described above relate to observed features of plutons, but they do not address the problem of why the magma rises. The magma came into existence because of the addition of heat to solid rock, presumably because the rock was carried by subduction or deformation to greater depth, into an environment that was normally hotter. The products of melting would be expected to have a lower specific gravity than the unmelted residue in the surrounding rocks. It may be for that reason that the force that caused the magma to rise was gravity; heavy rock pressing downward could cause less heavy magma to rise. This may seem fanciful but it will seem less so when we reflect on a geologic phenomenon not present in this part of the world but common elsewhere. This is the process of diapirism mentioned earlier in connection with serpentinite.

In the Gulf Coast region of Louisiana and Texas, and in many other places in the world, hundreds of bodies of rock salt, called salt domes, have about the same form as plutons. They are known in great detail because thousands of wells have been drilled into and around them for the reason that petroleum is associated with them. Even the interiors of a few can be examined in salt mines. They are roughly cylindrical, and vertically elongate. Sedimentary rocks close to them are bent upward around them, and have been faulted by the pressure of the salt. The sedimentary rocks over them are bent into a dome.

Most have diameters between 0.5 and 2 miles, but one is 7 miles in diameter. Many are 8 miles tall. The salt rises from a layer of sedimentary salt by perforating the overlying sedimentary rocks and pushing them both aside and upward. The salt flows plastically in the solid state. There can be no doubt that the driving force is gravity acting as a result of the fact that the specific gravity of the salt is less than that of the overlying sedimentary rock.

Thus it seems possible, by analogy, that the magma of plutons rises for the same reason that the salt rises, and that, like salt, it pushes aside and upward the rock that surrounds it. Unlike salt, magma can react chemically with the rock that it penetrates and become injected into joints, which provides additional possibilities for the making of room for itself.

Many minor aspects of plutons are interesting. Plutons have associated with them dikes that are unlike the granitic rock of the pluton, but which are derived from the same magma. As magma crystallizes, the ever-decreasing amount of remaining liquid changes in composition. Some of this liquid becomes separated from its parent body of crystallizing rock, then breaks out along fractures and intrudes both the parent and the surrounding metamorphic rock to form dikes of a fine- and even-grained light-colored rock called aplite. Dikes of aplite, as shown in fig. 77, are conspicuous along State Highway 70 in the Feather River Canyon near Grizzly Creek.

Some of the residual liquid from crystallizing magma becomes enriched in water, and in elements that are present in only very small amounts in the original magma. Some of this liquid may be entrapped locally in the plutonic rock to remain entirely enclosed by it, but some may be intruded along fractures into the parent plutonic rock or into the surrounding metamorphic rock. Because of the high content of water the resulting rock, called pegmatite, is characterized by very coarse grain size. Whereas the granitic rocks have grains of one-eighth to one-quarter inch, those of pegmatite are to be measured in inches or even in feet. Uncommon or rare minerals are present in some pegmatites, as are some semiprecious gemstones such as tourmaline and aquamarine.

Pegmatites are not common in this region, and the few that we have are not giant grained, and are not known to contain minerals of gem quality. Some rather badly weathered dikes of pegmatite only a few inches wide are present in the banks of the road from Portola to Lake Davis, just north of the cattle guard about 2 miles from Portola. A wider pegmatite dike is present in the road cut at the east end of the dam at Lake Davis. These dikes are all silicic; they contain quartz, feldspar, black mica (biotite), and a little black or very deep blue tourmaline.

Fig. 77. Dikes of aplite (light colored) in the mafic plutonic rock gabbro (dark colored). On State Highway 70 at the south end of the tunnel at Grizzly Creek, North Fork of the Feather River. The gabbro is a marginal part of the Grizzly Pluton.

Some unusual pegmatites of mafic composition, one of which is illustrated in fig. 78, are present in the metapyroxenite of Eureka Peak, which is not one of the granitic plutons related to the Nevadan Orogeny but is part of a much older one that has been metamorphosed. This is the pluton already mentioned as a possible source vent for the volcanic rocks of the Taylor or Goodhue formations. The pegmatites can be reached by the trail along the east side of Eureka Lake. There on the bare rock surfaces one can see roughly equidimensional bodies of pegmatite that consist of an outer zone of elongate crystals of augite, some as long as ten inches, radially arranged with respect to a core of fine-grained, light-colored rock. At the outer limit the coarse augite grades into the pyroxenite that has a grain size of about a half inch. Toward the center the large augite grains have crystal faces surrounded by the light-colored rock. From these relationships it is clear that the pegmatites are segregations of residual liquids from the pyroxenite when it was nearly crystallized, and that the pegmatite crystallized from the margin toward the center. The largest of the pegmatite segregations is about 20 feet in diameter but some are only a foot or so across.

The geologic ages of the metamorphosed sedimentary and vol-

Fig. 78. A segregation of pegmatite in the metapyroxenite of Eureka Peak, Plumas-Eureka State Park. The pyroxenite that surrounds the segregation, clearly visible in the lower right, is composed of grains of the mineral augite, about a quarter inch in diameter. The outer zone of the pegmatite consists of augite crystals as long as 6 inches that point toward the center. Their inner ends are partly surrounded by the light-colored rock of the central zone. The central zone consists mostly of fine-grained quartz and feldspar.

canic rocks described in Chapter 3 are relative ages based on the few fossils of marine organisms that have been found in them. The ages of the plutons, as based on their relationship to those fossiliferous rocks, are not closely determined. Here in the northern Sierra Nevada the youngest rocks intruded by the plutons are of late Jurassic age, about 150 million years old, and the oldest rocks that rest on the plutons are late Cretaceous in age, about 80 million years old.[2] Accordingly, the best that can be said about the age of the plutons, based on fossils, is that they are less than 150 million years and more than 80 million years old, which leaves them in a gap of uncertainty about 70 million years long.

However, plutons can be dated directly by radiometric methods that yield their ages in years. The methods, applied mostly to igneous rocks, involve the measurement in a sample of rock of the amounts

2. The ages of late Jurassic and late Cretaceous expressed in years—that is, their absolute ages—were not determined here, but at other places in the world where the circumstances for radiometric dating were favorable.

of certain radioactive elements and the elements into which they change. The age determination is possible because it is known that each radioactive element changes spontaneously at a constant and unalterable rate, expressed as its half-life. The half-life is the time in years during which half of the radioactive substance in any sample changes into a new substance called its decay product. Knowing the half-life in years, and the amounts of the radioactive element and its decay product in a sample, the time at which the decay process began can be calculated. That is presumed to be the age of the sample. In the case of igneous rocks, that is taken to be the time at which crystallization of the magma was completed. Although the amounts of material involved are very small, they can be measured with great precision by means of an instrument known as a mass spectrometer. Radioactive transformations useful in dating rocks involve certain isotopes[3] of uranium, thorium, strontium, potassium, and carbon and their decay products. Because potassium is present in most igneous rocks, and because the potassium dating process is relatively simple, it is the one most frequently used in age determinations. Only a small fraction of all the atoms of potassium in a rock are of the radioactive isotope of atomic weight 40, symbolized as ^{40}K, that decays into the isotope of the gas argon, symbolized as ^{40}Ar. The amount of ^{40}Ar in the sample must be determined. In practice it has been found necessary to determine only the total amount of potassium in the sample, instead of ^{40}K, because the ratio of the two is so nearly constant that the difference does not matter. From these two values and the known half-life of ^{40}K, the age is computed.

The method is impeccable, but there may be geological considerations that need to be taken into account. For example, it is not desirable to use as a sample a whole rock, but, instead to extract from the rock the crystalline mineral grains that contain potassium and use them as the sample. If one does this and makes determinations from one or two feldspars, from mica, and from hornblende, the "ages" of each may be somewhat different. There may be reason for averaging the values or for accepting one as better than the others. One reason for such differences is that some minerals are known to retain the argon better than others that permit some of the argon to "leak" away. Another geological consideration is that the argon may not have accumulated continuously since crystallization, but only since the rock had cooled to some lower temperature.

If an igneous rock is reheated—as, for example, by the emplacement of another intrusion nearby or by metamorphism—after once

3. Isotopes are varieties of atoms of the same element that have the same atomic number but different atomic weight.

having cooled, some or all of the accumulated argon may be released, and the age calculated is that of the time of reheating, or of some intermediate age if only part of the argon was released. It may be difficult or impossible to know whether a rock has been reheated, or to know how much argon was released if the rock had in fact been reheated.

If a magma has been contaminated by the incorporation into it of older rocks from the walls that bound the magma it may acquire added amounts of potassium and ^{40}Ar. One may assume that all of the ^{40}Ar in the older rock was released by the melting or dissolution of the older rock, in which case the age of the igneous rock that resulted from crystallization of the magma would be unaffected, but there must always be uncertainty; the age may be too old or too young. It is well to avoid dating rocks known to have been contaminated. Some of these problems can be avoided by using other radioactive elements but the processes are more costly and time consuming.

The absolute ages of many plutons in the Sierra Nevada have now been determined by several different methods. They range between 210 and 80 million years and seem to fall into five groups. Although most of the dates are of rocks in the central part of the range, a few are from the north. Samples from both the Bucks Lake and Merrimac plutons (see fig. 75), are about 130 million years, or early Cretaceous. A few dates from the eastern plutons in Plumas and Lassen counties fall in the range of 90 to 120 million years, and one sample from the Granite Range near Gerlach, Nevada, is 89 million years old, all of which dates are Cretaceous. It seems that the plutons in the western part of the northern Sierra Nevada are older than those farther east. In the southern Sierra Nevada there is a well-established progression from older to younger eastward in the range among the plutons of later Jurassic and Cretaceous age, but older plutons in the age range of 210 to 160 million years are present in the southern Sierra Nevada and the Inyo Mountains.

In the north, then, the oldest plutons, about 130 million years old, seem to have been emplaced not long after the beginning of the Nevadan Orogeny, about 140 million years ago. Younger plutons farther east, about 90 million years old, are as young as the unmetamorphosed fossiliferous sedimentary rocks of Cretaceous age that rest unconformably on the 128-million-year-old Rocklin Pluton east of Sacramento.

So present information indicates that the invasion of magma to form plutons extended not continuously, but as a series of episodes or pulses through a period of about 55 million years, considerably longer than the deformational part of the orogeny, which may have been completed in about 22 million years.

For the most part the plutons seem to be younger than the cleavage of the metamorphic rocks that they intrude, but some plutons were deformed. It has been suggested that some of the plutons, especially the youngest ones, broke through to the surface to produce volcanoes. But no volcanic rocks known to be of that age have yet been found on either side of the northern Sierra Nevada that would support such an idea.

Termination of the episodes of intrusion of plutons marks the end of the Nevadan Orogeny.

This history of the Ancestral Sierra Nevada lacks many details, partly because, of necessity, this must be an abbreviated story, but partly also because much remains to be discovered. Neither is the history exactly the same throughout the entire length of the terrain. For example, rocks as young as Late Cretaceous are involved in the deformation both to the north and south of California. However, in spite of variations, it is true that a belt of metamorphic rock invaded by plutons was added to the western margin of the North American continent in the later Jurassic and Cretaceous periods. This was the Ancestral Sierra Nevada, a vastly larger terrain than that which we now call Sierra Nevada. On the south the typical rocks can be traced across the Mojave Desert and along the Peninsular Range and far into Mexico. On the north the belt of rocks passes under younger rocks into the Klamath Mountains, where it turns to the northeast, passes beneath the southern Cascade Range and the plateau to the east, and reappears in the Blue and Wallowa mountains of eastern Oregon. From there the belt trends northwesterly under the young volcanic rocks of eastern Washington to reappear once more in the northern Cascade Range of Washington, whence it parallels the Pacific Coast into southeastern Alaska (see fig. 39). The Nevadan Orogeny was a great event in the earth's history and, without doubt, it produced a great, mountainous land.

However, during the next 40 million years in California the range was subject to erosion that reduced whatever mountainous elevations that were once present to a relatively featureless plain—a land of low relief. After that lapse of time, during which we know almost nothing about the region, the sea once again lapped on the western slope of what we now call the Sierra Nevada, and the thread of geologic history written in the rocks can be picked up once again.

PART III

The Sierra Nevada After the Nevadan Orogeny

6

The Rocks of the Superjacent Series

INTRODUCTION

During the span of some 20 million years after the Nevadan Orogeny, erosion had reduced the Ancestral Sierra Nevada in northern California to a lowland plain. The evidence for this is of two kinds. First, the products of that erosion were deposited beneath the present Sacramento Valley and are well known through the drilling of a multitude of wells in search of petroleum and natural gas. The deposits are essentially of mud and sand with but few pebbles. They are very unlike the beds of boulders and coarse gravel now visible in the river beds and in the gold-dredging grounds near Oroville, Marysville, and Folsom that are the products of erosion of the present rugged range. From that fact one infers that whatever elevations comparable to the present that might have existed had been eroded away by the time that deposition was renewed. Second, the surface of unconformity on which the rocks of the Superjacent Series were deposited, which was also the result of that erosion, can be observed; such direct observation indicates a surface of low relief.

At the time that the geologic history of the northern Sierra Nevada once more began to be recorded in rock, the Pacific shoreline lay close to what is now the western edge of the range. Although the position of the shore undoubtedly fluctuated with time, it never again crossed the site of the Sierra Nevada. The Chico, "Dry Creek," and Wheatland formations are marine; they were deposited in the margin of that sea, and they belong more to the province of the Sacramento Valley than to the Sierra Nevada. Excepting them and a few marine layers in the Ione Formation, the rocks of the Superjacent Series were all deposited above sea level. Their present distribution is shown on figs. 79 and 80.

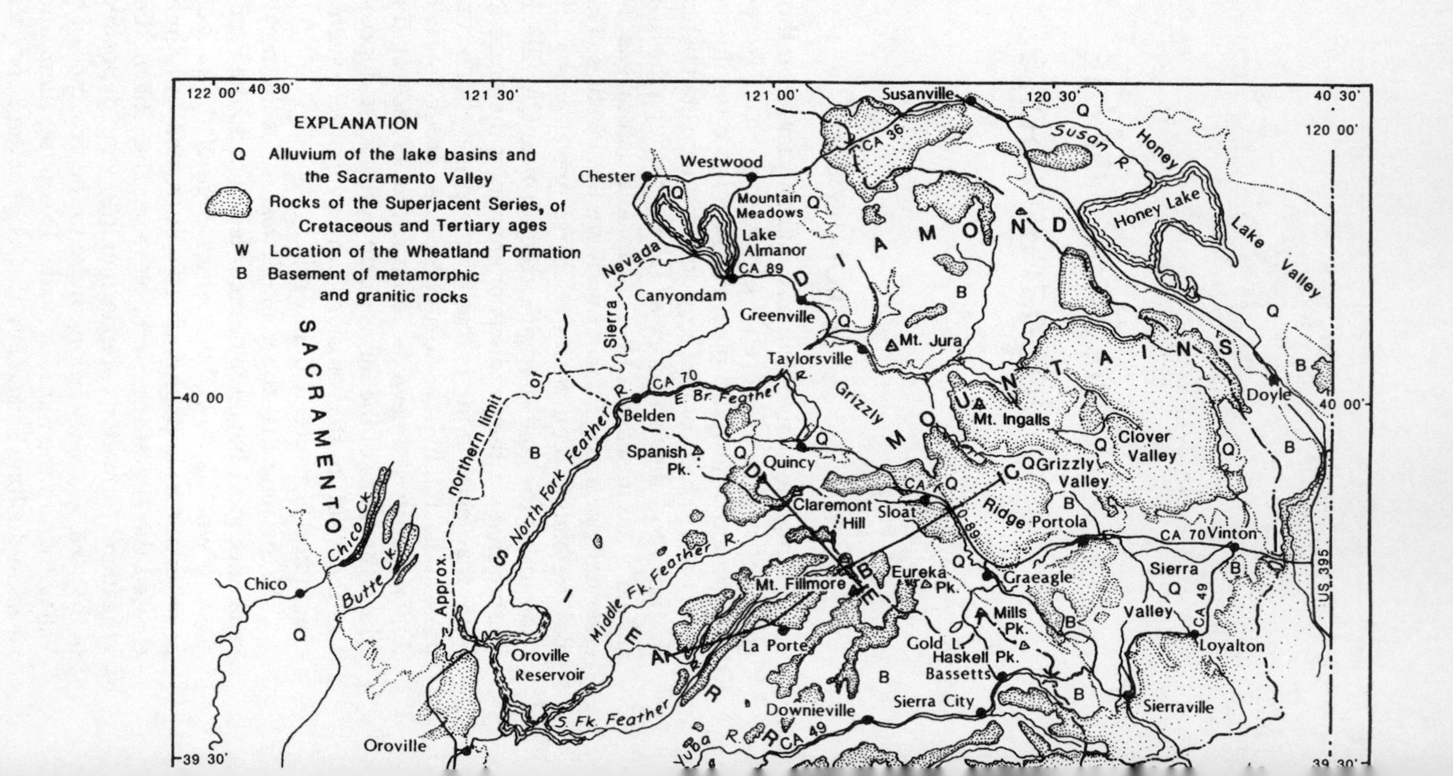

EXPLANATION
Q Alluvium of the lake basins and the Sacramento Valley
Rocks of the Superjacent Series, of Cretaceous and Tertiary ages
W Location of the Wheatland Formation
B Basement of metamorphic and granitic rocks
122 00'
40 30'
121 30'
121 00'
120 30'
120 00'
40 00
40 00'
39 30'
SACRAMENTO
DIAMOND MOUNTAINS
GRIZZLY MOUNTAINS
SIERRA NEVADA
Approx. northern limit of Sierra Nevada
Susanville
Westwood
Chester
Mountain Meadows
Lake Almanor
Canyondam
Greenville
Mt. Jura
Taylorsville
Belden
Spanish Pk.
Quincy
Claremont Hill
Sloat
Mt. Ingalls
Clover Valley
Grizzly Valley
Grizzly Ridge
Portola
Vinton
Doyle
Honey Lake
Honey Lake Valley
Susan R.
Sierra Valley
Loyalton
Sierraville
Graeagle
Eureka Pk.
Mills Pk.
Gold L.
Haskell Pk.
Bassetts
Sierra City
Downieville
La Porte
Mt. Fillmore
Oroville Reservoir
Oroville
Chico
Chico Ck.
Butte Ck.
North Fork Feather R.
E. Br. Feather R.
Middle Fk. Feather R.
S. Fk. Feather
Yuba R.
CA 36
CA 89
CA 70
CA 49
US 395

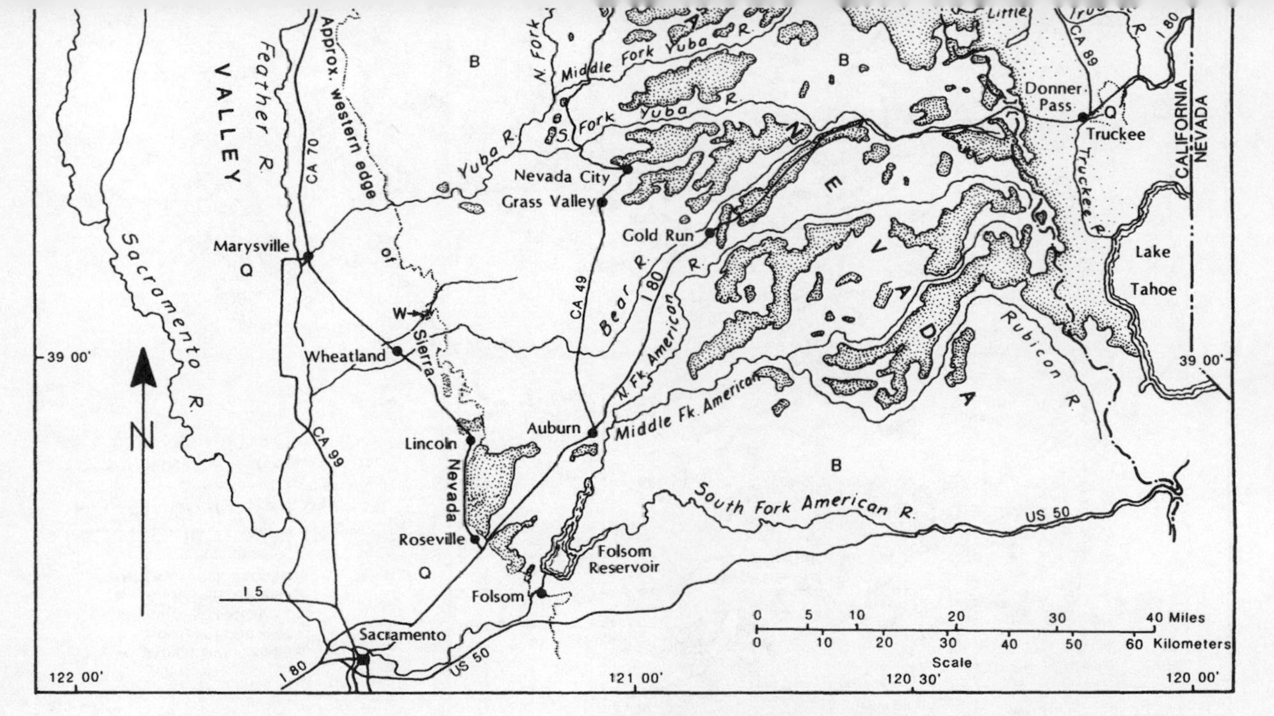

Fig. 79. Map of the northern Sierra Nevada and Diamond Mountains showing the distribution of all the formations of the Superjacent Series combined. The line A–B–C indicates the position of the geologic cross section A–B–C of fig. 129, and the line D–E indicates the position of the profile D–E of fig. 129.

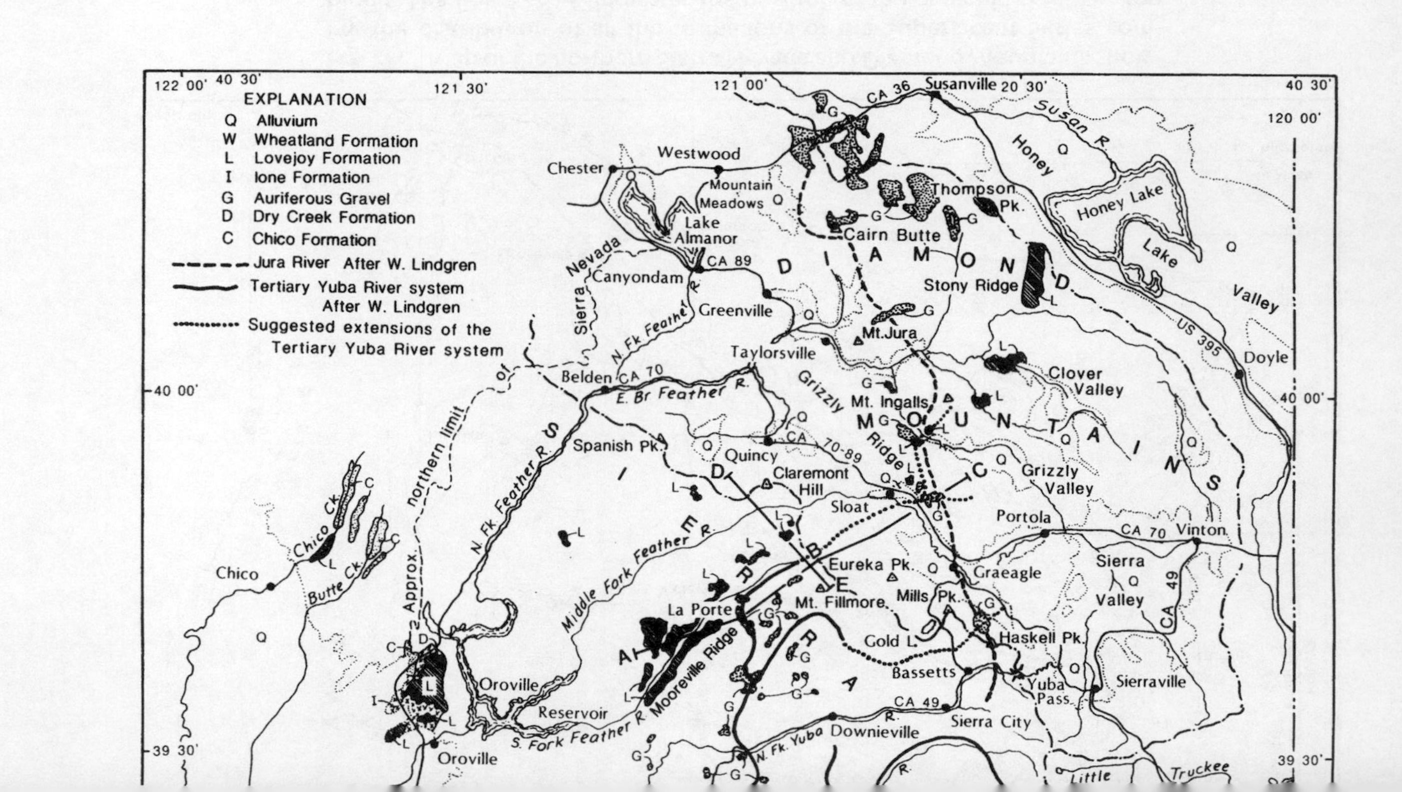

EXPLANATION
Q Alluvium
W Wheatland Formation
L Lovejoy Formation
I Ione Formation
G Auriferous Gravel
D Dry Creek Formation
C Chico Formation
Jura River After W. Lindgren
Tertiary Yuba River system After W. Lindgren
Suggested extensions of the Tertiary Yuba River system
122 00'
40 30'
121 30'
121 00'
20 30'
120 00'
40 00'
39 30'
Susanville
CA 36
Susan R.
Honey
Honey Lake
Lake
Valley
Westwood
Chester
Mountain Meadows
Lake Almanor
Thompson Pk.
Cairn Butte
Stony Ridge
CA 89
Canyondam
Sierra Nevada
Greenville
N. Fk. Feather
Mt. Jura
Taylorsville
US 395
Doyle
Clover Valley
Belden
CA 70
E. Br. Feather R.
Grizzly
Mt. Ingalls
Ridge
Spanish Pk.
Quincy
CA 70-89
Claremont Hill
Sloat
Grizzly Valley
Portola
Vinton
DIAMOND MOUNTAINS
SIERRA
Chico
Chico Ck.
Butte Ck.
Approx. northern limit of
N. Fk. Feather R.
Middle Fork Feather R.
Eureka Pk.
Graeagle
Sierra Valley
CA 49
La Porte
Mt. Fillmore
Mills Pk.
Gold L.
Haskell Pk.
Bassetts
Yuba Pass
Sierraville
Oroville
Reservoir
S. Fork Feather R.
Mooreville Ridge
Sierra City
Downieville
N. Fk. Yuba
Little
Truckee

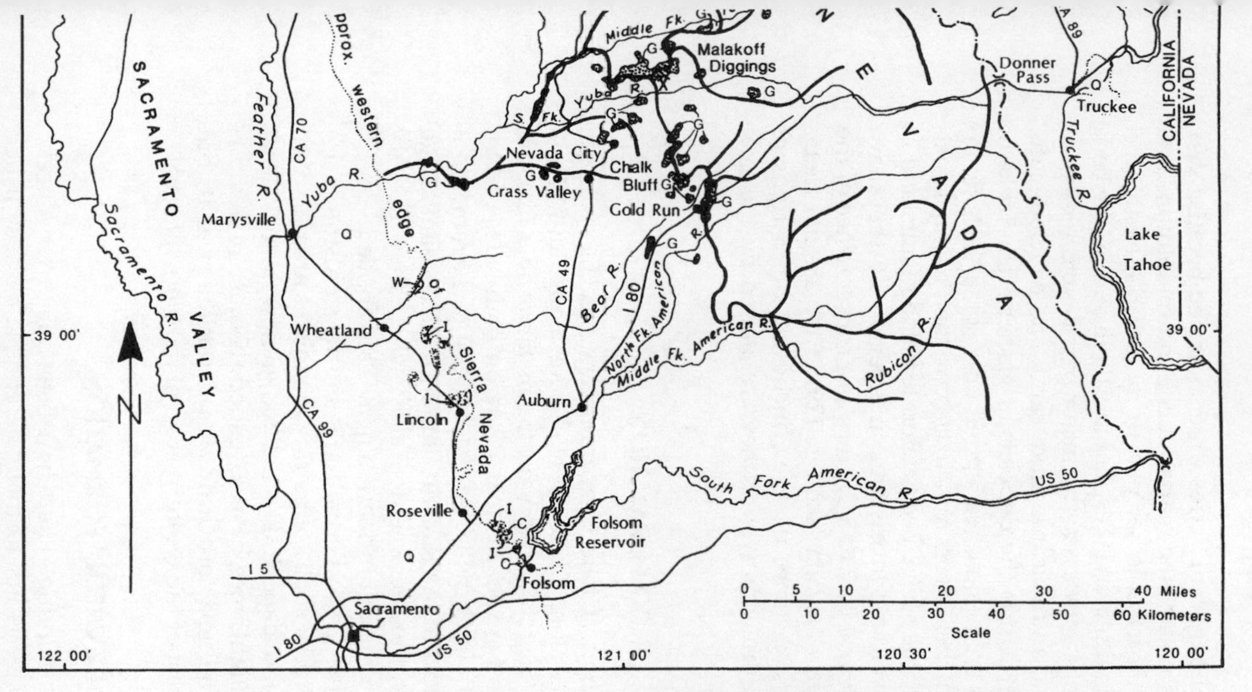

Fig. 80. Map of the northern Sierra Nevada and Diamond Mountains showing the location of the older formations of the Superjacent Series, and the courses of the early Tertiary (Eocene) rivers according to W. Lindgren. Possible northeasterly extensions of the Tertiary Yuba River system are indicated.

The rocks of the Superjacent Series were so named because they rest on top of the metamorphosed marine rocks and the plutons that invaded them. They consist in lesser part of mud, sand, and gravel deposited in the marginal sea, in lakes, but mostly in and on the beds and floodplains of rivers. However, the major part is volcanic—lava flows, mudflow breccia, and volcanic mud, sand, and conglomerate. A minor but important part is the deposits of glaciers. The rocks of the Superjacent Series are layered, excepting only some volcanic intrusions, and in general their layers remain close to their original position. Locally, however, they have been deformed along faults and at the margins of some volcanic intrusions, and regionally they have been tilted to the southwest by 2 to 3 degrees as a consequence of the recent uplift of the Sierra Nevada (see fig. 129). They are not metamorphosed except for a distance of a few inches at the margins of some of the volcanic intrusions where they are altered by the heat from the adjacent magma.

CHICO FORMATION

The Chico Formation of Late Cretaceous age (see fig. 80) was named for its occurrence along Chico Creek. There, along Butte Creek and elsewhere, it rests unconformably on the metamorphic rocks. It is also present but poorly exposed near Pentz on both sides of Dry Creek a few miles north of Oroville. At Folsom east of Sacramento it rests on the granitic rocks of the Rocklin Pluton. Rocks of similar age are continuous beneath the Sacramento Valley, where they are very thick and where many different formation names are applied.

In Butte County, the Chico Formation is mostly sandstone with some beds of pebbles; fossil shells of marine organisms are plentiful. It is definitely a near-shore deposit. It is not known how far to the east the Chico Formation might once have extended, because it was certainly eroded before the next formation was deposited over it, but none has been found east of the western edge of the Sierra Nevada.

"DRY CREEK" FORMATION

Although the rocks mentioned here have been called "Dry Creek" for a long time, the name is not formally acceptable because it had been applied previously to rocks elsewhere, and because geologists adhere to the rule, among others established by themselves, that formation names should not be duplicated in North America. Because the rocks have not been formally renamed, "Dry Creek" is put in quotation marks.

The "Dry Creek" Formation overlies the Chico Formation on the south side of Dry Creek north of Oroville (see fig. 80) and evidently extends southward underneath Oroville Table Mountain. It contains marine fossils that show it was deposited during the middle of the Eocene epoch. The fact that a great thickness of rocks of Paleocene and early Eocene age that are present under the central Sacramento Valley are not present here leads to the inference that the Chico Formation was eroded before the "Dry Creek" Formation was deposited on it.

Although the "Dry Creek" Formation is very restricted in occurrence and poorly exposed, enough can be seen to show that it is sandstone deposited near a shoreline and that, like the Chico Formation, it indicates that during its time of deposition the Sierra Nevada was a land of low relief compared to its present state.

IONE FORMATION

Named for the town of Ione in Amador County, the Ione Formation is well known there because it contains low-grade coal (lignite) that was mined in earlier times and commercially valuable clays that have been produced for many decades. But the Ione is mostly gravel and sandstone rich in quartz. The formation rests on a deeply weathered surface of the metamorphic and granitic rocks of the Sierra Nevada. These features have been taken to indicate that at the time the Ione was deposited the Sierra Nevada was a land of low relief with a humid subtropical or tropical climate, matters that are discussed in more detail in connection with the Auriferous Gravel. The formation is present at the surface discontinuously along the western margin of the Sierra Nevada from Friant on the San Joaquin River to Oroville Table Mountain (see fig. 88), where it lies beneath the black cliff-forming lava of the Lovejoy Formation. Near Oroville it consists mostly of sandstone with abundant quartz. But scattered in the sandstone are well-rounded pebbles and cobbles of the volcanic rock hornblende andesite.

The age of the Ione is established as middle Eocene by marine fossils found in it in a few places and by the fact that it grades westward under the Great Valley into the marine Domengine Formation of that age. To the east it is continuous with the gold-bearing gravels described next. These relationships permit the conclusion that the Ione was deposited by westward-flowing streams at the edge of the present Sierra Nevada and in the adjacent shallow sea—it is a shoreline deposit. No doubt the coal represents the accumulation of vegetation in coastal lagoons behind barrier bars such as are common along modern coastlines.

AURIFEROUS GRAVEL

Aside from the value of the gold recovered from it, the Auriferous Gravel is one of the most fascinating rock units of the Sierra Nevada. The name means simply gold-bearing gravel, and although it is not properly a formation name—because there are other gravels of different age and character that also contain gold—the name is so well established that local geologists understand its use in the sense of a formation. The term includes not only gravel but beds of boulders, sand, and clay that are of middle Eocene age. Most deposits are present on the divides between the deep canyons of the modern rivers, where they rest on deeply weathered metamorphic and granitic rocks.

Gold was first discovered in central California in the South Fork of the American River at Sutters Mill, near Coloma, on January 24, 1848, by James W. Marshall. The Gold Rush began as soon as the discovery became public knowledge, and men came in unprecedented numbers to seek their fortunes extracting gold from the beds of the great rivers. Mostly they were not miners but enthusiastic amateurs, who, not knowing where gold ought to be found, looked for it everywhere. They found gold also in the tributaries of the major rivers, and, working ever higher upstream, they were led to the broad plateaulike divides between the deep canyons where they found gold not only in the beds of flowing streams, but also in gravel not related to the many modern streams. It was soon recognized that this was old gravel, preserved in patches, some of it extending beneath younger rocks, and that it was the relict of ancient river deposits. The ancient river beds were called channels by the miners, and the gravel and associated sand and clay became known as the Auriferous Gravel. At first thought to be of Miocene age, it was later recognized to be middle Eocene, 50 million years old, more or less.

Hydraulic mining was invented to exploit the gravel, and during the time that the method was in use many miles of the ancient river beds were laid bare (fig. 81). Those parts that were too deeply buried to be profitably worked by the hydraulic method were mined underground, so that scarcely any part remains unexploited.

The Auriferous Gravel filled valleys with gently sloping sides to a width of as much as 2 miles and to a thickness, in at least one place, of more than 400 feet. Below the gravel there is an inner or deepest channel cut into the underlying older rock, in which the stream flowed just before the bulk of the gravel began to accumulate. In most places the inner channel is only a few feet deep and a few tens of feet wide, but the gravel that filled it, which was very coarse and composed of rocks of obviously local origin, contained most of the gold. Finely divided pyrite that formed in the gravel of the inner channel caused it to

Fig. 81. The Auriferous Gravel. A view downstream along an excavated part of the La Porte channel between Gibsonville townsite and Whiskey Diggings. The boulders on the floor of this old river bed are only the largest of those in the gravel. They were left behind by the miners using the hydraulic mining method. The bank in the lower left corner of the picture is unmined gravel rich in quartz pebbles about an inch in diameter in a matrix of quartz sand grains and clay. SE 1/4, Sec. 19, T. 22 N., R. 10 E.

have a bluish or bluish green color. For that reason the miners called it blue gravel or blue ground and many channels were named the Blue Lead.[1] They also called it deep gravel, and the name Deep Blue Lead was also commonly applied.

As more gravel accumulated, it spread over the gently sloping valley sides that miners came to call benches. They called the gravel there bench gravel, or, if they found that it contained little or no gold, which was often the case, they called it bastard gravel.

Beds of sand and clay commonly accompany the gravel; the assemblage is one to be expected of aggrading streams, that is, streams that are building up their beds by depositing layers of sediment on their floodplains. Such streams change position on their floodplains; they cut new channels and backfill old ones; they isolate parts of old

1. Pronounced as in "leader."

channels that then become ponds that later fill with clay and silt. Such processes result in steeply inclined and curved cross beds, in filled channels, and in rapid changes in grain size both horizontally and vertically. All of these are features of the Auriferous Gravel, and they and the old channel bottoms can be seen in the now long unworked mine pits at such places as Gibsonville, La Porte, Howland Flat, St. Louis, Port Wine, and Scales. Perhaps the best place of all to see the nature of the gravel is at Gold Run in the great cliff on the north side of Interstate Highway 80 east of the rest stop.

However, the blue gravel is not to be seen. Very little was left unmined, and what was left changes color on exposure to air because the pyrite weathers quickly to brown or rusty red iron oxide. Furthermore, the deepest parts of the old pits have now mostly been at least partly filled with material eroded from the pit walls during the many decades since hydraulic mining was effectively stopped first by an injunction issued by the U.S. Circuit Court, and later by an act of Congress, both of which prevented the unregulated dumping of mining debris into the rivers.

Typically, the bench gravel is light yellow or pinkish owing to iron oxide, and the overall aspect is lightness of color. Even a casual inspection reveals that white quartz composes a large proportion of the cobbles, pebbles, and sand grains, but clay is also an important constituent of both the sand and gravel. The men who first studied the Auriferous Gravel thought that it, excepting only blue gravel, consisted almost exclusively of minerals that did not readily decay in weathering—quartz and gold—and that these two minerals were derived from veins in the older metamorphic rocks east of the gold-rich gravel that was found mostly in the middle reaches of the range. This idea, together with the knowledge that older rocks beneath the Auriferous Gravel are deeply decayed, and that the sands of the Auriferous Gravel do contain but few minerals that are resistant to weathering, other than quartz and gold, led to the notion that the climate under which the materials of the gravel came into existence was tropical. However, it is now known that not all the pebbles were resistant to weathering, that not all of the white pebbles were vein quartz (fig. 82), that not all the sand grains were quartz, that some of the gold may have come from Nevada, and that the climate was not truly tropical.

Later studies show that as much as half of the white and nearly white pebbles in some places are chert and quartzite, which are as resistant to weathering as is vein quartz. In other samples as much as a third of the original pebbles and cobbles are now clay and some of the larger of these have cores that are sufficiently well preserved to reveal their original character. Many of them are not rich in quartz and are

Fig. 82. The Auriferous Gravel. Cobbles at Howland Flat left behind by the hydraulic mining operation. Note that only a few are very white—that is, composed of quartz from veins. The rocks in various shades of gray are mostly chert.

not notably resistant to decay. The conversion to clay is the result of weathering, or decomposition, of the gravel in place—that is, after it was deposited. The same is true of the sand. Much sand is very rich in quartz, but not all the quartz was derived from veins. Clay derived in place from weathered sand grains is as much as 40 percent of some samples. Much of the clay came from feldspar, for easily weatherable but undecayed feldspar minerals make up more than half of some samples of sand. The mineral anauxite, a variety of the common clay mineral kaolin, is rather conspicuous in both sand and gravel. It occurs as pearly white flakes, many of which are sharply hexagonal in outline and about sand grain size. Because it is a very soft mineral it cannot have survived transportation with the hard sand grains of quartz and feldspar, much less with the pebbles of the gravel. It must have originated in place after deposition. It is thought to have originated through the decomposition of the black mica biotite, which it closely resembles in form. Accordingly, the presence of the unaltered cores of cobbles, of unaltered feldspar, and of the anauxite precludes the notion that the Auriferous Gravel was the unalterable residue of a terrain subject to weathering in a tropical climate. A climate like that of the southeastern United States could produce the required results.

The same conclusion applies to the Ione Formation, since it is physically continuous with the Auriferous Gravel.

Fossil leaves preserved in layers of clay in the Auriferous Gravel provide a better approach to the interpretation of the climate. They have been found in a number of places in the northern Sierra Nevada, but those at Chalk Bluff (see fig. 80) in Nevada County have, through the studies of MacGinitie, yielded the most information. The leaves are mostly those of forest trees. The forms of the leaves, their texture, patterns of venation, the assemblage of species, and other pertinent matters led MacGinitie to the conclusion that the climate was like that at Orizaba, Mexico, which is on the eastern slope of the mountains at 19 degrees north latitude. The average annual temperature there is about 65° F, there is no frost, and the average annual rainfall is 60 to 80 inches, with a warmer wet season and a cooler dry season. This is subtropical, not tropical. It is an unvarying or equable climate, and such an equable climate seems to have extended over much of North America in middle Eocene time. Why the climate then was less strongly zoned than it is now is an as yet unresolved problem, but as is indicated in the discussion of the Ice Age of the Quaternary period, climate surely has changed with time.

In view of the now well-established facts of plate tectonics and continental (or plate) drift, one may ask if California did not have a subtropical climate in middle Eocene time because the region then occupied a more nearly equatorial position than it does at present. The answer indicated by the study of the magnetism of rocks is that it did not. At the time of formation of many kinds of rocks, the direction of the earth's magnetic lines of force existing at that time is preserved in them. By determining this "fossil" magnetic direction in rocks of the same age at many places over the continent the position of an apparent magnetic pole for that age can be established. This determination has been made for rocks of many different ages and has provided an apparent path of wandering of the magnetic pole. If the real magnetic pole has not moved, the continent must have moved and the apparent wandering path of the pole is a real wandering path of the continent. This is not, of course, quite such a simple matter, but fairly satisfactory paths have been found for the several continents. From present information it seems that since early Eocene time, the latitude of California has not changed enough to account for any such climate as existed here in middle Eocene time.

Fossil leaves of the same significance as those at Chalk Bluff are known farther north at Cherokee, north of Oroville; in the Diamond Mountains near Cairn Butte on the ridge about 13 miles due north of Taylorsville; and at a now lost locality described as 7.5 miles southwest of Susanville. Fossil leaves also occur at La Porte on the west side

Fig. 83. The Auriferous Gravel. The wall of the Upper Dutch Diggings hydraulic mine at La Porte. The bed that contains fossil leaves is at the top, and a hardened portion of it forms the outstanding ledge in the upper left corner of the picture. The remainder of the bank consists of clay with sandy and gravelly layers. SE 1/4, Sec. 8, T. 21 N., R. 9 E.

of the Upper Dutch Diggings (fig. 83). There the gold-bearing gravel is overlain by beds of clay and sand and those in turn are overlain by bedded volcanic ash that contains the leaves. The thinly layered ash was obviously transported by water. As the ash is not a part of the Auriferous Gravel, the leaves are believed to indicate a slightly younger age, as does a radiometric date of the ash of 29 million years before the present, or middle Oligocene. The leaves are of such species and character as to indicate that the climate then was essentially like that inferred from the leaves at Chalk Bluff.

The geologists who first studied the Sierra Nevada, J. D. Whitney, J. S. Diller, H. W. Turner, and W. Lindgren, were able to demonstrate that the Auriferous Gravel north of the Stanislaus River is composed of the relict deposits of an extensive system of rivers that flowed westward but on courses more sinuous than those of the present rivers, and that all the ancient rivers now within the basins of the American, Bear, and Yuba rivers were parts of a large system that came to be known as the Tertiary Yuba River (see fig. 80). The southern and central branches of this system were believed to head at the present divide west of Lake Tahoe near Donner Pass and along the divide northwest of Donner Pass. This divide was believed to continue northwest at least as far as Onion Valley in Plumas County, where it is known as Bunker Hill Ridge. It was also believed that a stream they called the Jura River (see fig. 80), which headed near Sierra City, flowed northwest, parallel to but east of the divide.

It is well established that a segment of the Tertiary Yuba River in the central Sierra Nevada flowed northwest from near Forest Hill to North Columbia, where it was joined by other west- and southwest-flowing tributaries, and where it turned west and left the Sierra Nevada near Marysville. The southwest-flowing branches are the La Porte and Port Wine channels of Plumas and Sierra counties. Both disappear upstream under younger rocks, the first above Whiskey Diggings, under Bunker Hill Ridge, the other at Howland Flat, under Port Wine Ridge and Mt. Fillmore. No continuation of either was identified on the northeast side of the present divide in the drainage basin of Nelson Creek. It was these facts, among others, that led to the belief that the ancient divide at the head of the La Porte and Port Wine channels was at or close to the present one, and that there was a Jura River beyond the divide.

Several reasons why the La Porte and Port Wine channels appeared to end there are now known. Absence of the gravel is in part the consequence of erosion. Episodes of erosion occurred not just once but several times before the present cover, the rocks of the Penman Formation, were deposited over the gravel. Large areas of geologically recent landslides on both sides of Bunker Hill Ridge prevent direct observation of the La Porte channel, but geological studies of the area by R. L. Strand show that it is necessary to conclude that the channel is cut off under a landslide by a fault that probably is older than the Penman Formation. Many younger intrusions of the volcanic rock andesite have removed large parts of the channels. Pilot Peak, Mt. Stafford, Mt. Etna, Beartrap Mountain, and Blue Nose Ridge are all of the younger intrusive igneous rock.

The nature of the channel fill also precludes the possibility that the present crest was the headwaters. A considerable part of the gravel at La Porte, Whiskey Diggings, and Howland Flat is composed of varieties of chert, metamorphosed quartz sandstone, and metamorphosed volcanic rocks unlike any among the metamorphic rocks that are yet known farther east, northeast, or north. These are the rocks described in Chapter 3 that comprise a belt that extends from the North Fork of the American River to the region east of Lake Almanor. East of this belt is a vast area of granitic plutons that extends into Nevada. It appears that the chert, quartzite, and volcanic rocks in the gravel must have come from east of the plutons. The source region has not been more closely identified.

At both Howland Flat and above La Porte the Auriferous Gravel was described by Turner and Lindgren during the period of mining as including thick beds of clay. The deposits of clay in the La Porte channel above Whiskey Diggings are reported to have been more than 1,000 feet wide and more than 300 feet thick. Such a large channel fill and so much fine-grained clay are hardly to be expected at the headwaters of a river in a mountainous region!

Deposits of gravel are present farther north and northeast of the supposed headwaters of the La Porte and Port Wine channels, but they were assigned to the Jura River. The Jura River was supposed to have flowed from headwaters near Sierra City along a course marked by gravel at Haskell Peak, which is about 10 miles southeast of Johnsville, past gold-bearing gravel along the face of the mountain above the Mohawk Valley between Haskell Peak and Mills Peak, then to gold-bearing gravel high in the ridges on both sides of Little Long Valley Creek north of Cromberg. From there it would have crossed what is now Grizzly Ridge to the gravel at the Cascade Hydraulic Mine near the now inactive Walker Mine. Its supposed course went from there to gravel on the ridge north of Genesee, past gravel near Kettle Rock to Mountain Meadows Reservoir, where it was thought to turn eastward. Between Mountain Meadows Reservoir and Susanville there is a large area of gravel that was suggested to be part of a delta deposited in a basin into which the river flowed.

There are some problems about the concept of the Jura River. The gravel at Haskell Peak, and between there and Mills Peak, seems not to have come from the region around Sierra City or south of there. This is especially true of cobbles of pink granite. Also, although not well dated in that area, some of the gravel possibly belongs to the much younger Delleker Formation. Much of the gravel at Little Long Valley Creek (fig. 84) and at the Cascade Mine came from the north, from the rocks of the Taylorsville District. Another part probably came from northwestern Nevada. Cross-bedding indicates that the stream flowed southwesterly. This gravel is not as rich in quartz as that at Whiskey Diggings and Howland Flat, and although it is not accurately dated it may be a little younger. Clearly, transport was to the south or southwest, not to the northwest. Gravels farther north have not been studied in any detail, but they are not rich in quartz, and they do contain much rock of volcanic origin. It is these deposits at Cairn Butte and southwest of Susanville that contain fossil leaves determined to be Eocene in age. The direction of transport there has not been established, but it seems not to have been to the northwest because appropriate kinds of source rocks are not known to the southeast.

It thus seems quite clear that the Jura River never existed, and that the present northwest-trending divide around the head of the Yuba River was not a divide when the Auriferous Gravel was depos-

Fig. 84. The Auriferous Gravel. On Grizzly Ridge on the west side of the canyon of Little Long Valley Creek, center of Sec. 1, T. 23 N., R. 11 E. On the forest road from State Highway 70-89 at Rattlesnake Creek to Happy Valley. Cross-bedding here shows that currents flowed to the left, which is southwesterly. The cobbles are easily recognized as having been derived from the formations of Paleozoic age of the Taylorsville District not less than 5 miles to the northeast. Accordingly, the rocks have been transported to the south or southwest.

ited. Streams farther south that deposited Auriferous Gravel also had headwaters in Nevada, as indicated by the presence of gravel near Donner Pass and on the Carson Range east of Lake Tahoe.

The conclusion to be drawn from this discussion is that during middle Eocene time there was no Sierra Nevada that in any way resembled the present mountain range. Yet it is possible to say something about the nature of the landscape upon which the Auriferous Gravel was deposited.

The geologists who first described the Auriferous Gravel concluded that in the central Sierra Nevada there were gravel-filled canyons that were hundreds of feet deep—in places more than a thousand feet deep. They concluded also that certain ridges and peaks of rocks resistant to erosion stood above the alluvial plain formed by the gravel. They thought that the landscape of that time resembled that of the present. Some of their observations are undoubtedly correct, but it is also true that in the western slope of the Sierra Nevada there are many faults younger than the Auriferous Gravel that went unrecognized even though faults were exposed and described as such in the mines. Faults can change the apparent relationship between gravel and basement so that an unrecognized fault might pass for a buried slope. Moreover, there seems to have been no high relief along the channels in Plumas County. Figure 85 is a topographic map drawn on the surface on which Auriferous Gravel of the Port Wine and La Porte channels rests, in the region between La Porte, Port Wine, Gibsonville, and Howland Flat. The map, constructed by T. H. Hemborg, is not complete because the La Porte channel between La Porte and Gibsonville is still concealed beneath younger rock although the gold was removed long ago by underground mining. The position of the unconformity at the base of the younger Penman Formation (not shown on fig. 85) limits the upward extent of the contoured surface.

The maximum relief of the surface beneath the Auriferous Gravel shown on fig. 85 is about 600 feet near Howland Flat. The steepness of slope is reflected in the spacing of contours—the closer together, the steeper the slope. A feeling for slope can be obtained by comparing fig. 85 with fig. 86, which is a map of the same area as it is today, at the same contour interval. Clearly the surface beneath the Auriferous Gravel is less steep than the rounded upper slopes of Gibsonville Ridge. No part of the area has steep slopes like those of the inner gorge of Slate Creek, which is quite typical of the modern canyons of the Sierra Nevada.

The average gradient of the ancient streams is now about 100 feet per mile, but this tells us nothing about the landscape because the original gradient cannot now be determined. It would be a most unlikely and remarkable accident if it were the same as it is now because two kinds of disturbances have affected it, one of which is faulting. There are many faults in the region, only one of which is shown in fig. 85. Others are known on the surface and more were recorded in the underground mining of the La Porte channel but their precise positions are unknown. The movement on the faults is such as to make the present gradient appear to be less than it was in past time. The second kind of disturbance is the westward tilt of the Sierra Nevada, which makes the gradient appear steeper than it was in the past. Not enough is known about either faulting or tilting to permit a determination of the original gradient of the channels.

In summary, then, the landscape was at first one of rolling hills between broad valleys in which large rivers flowed along courses more sinuous than those of most of the present rivers. The local relief may have been as much as 1,000 feet, but probably it was much less. As gravel accumulated in the river valleys, floodplains developed

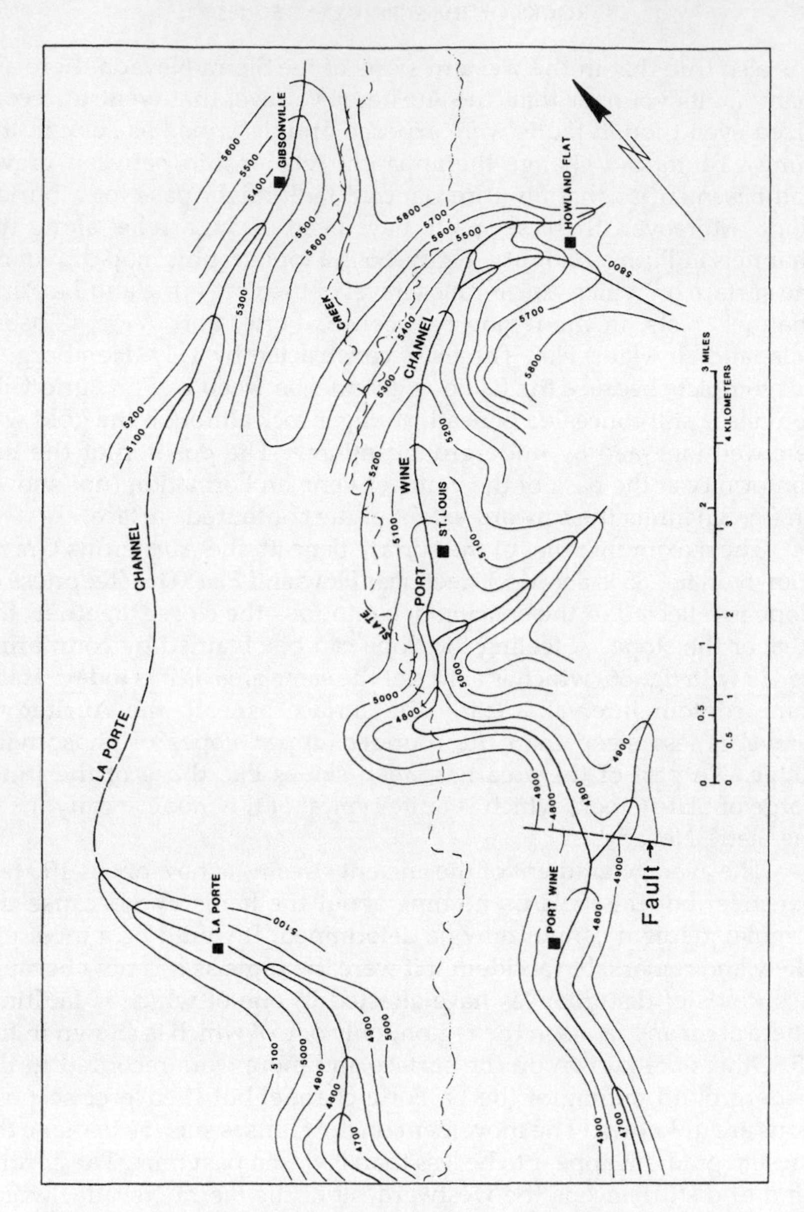

Fig. 85. A middle Eocene landscape. A topographic map of a part of the surface upon which the Auriferous Gravel was deposited in the La Porte–Gibsonville District. The map is incomplete because parts of the channels remain concealed beneath the younger rocks of the Penman Formation, and parts were lost to erosion. The course of Slate Creek is as it is today. Compare this to the present landscape of the same area shown in fig. 86 at the same scale and contour interval. From the unpublished studies of T. H. Hemborg.

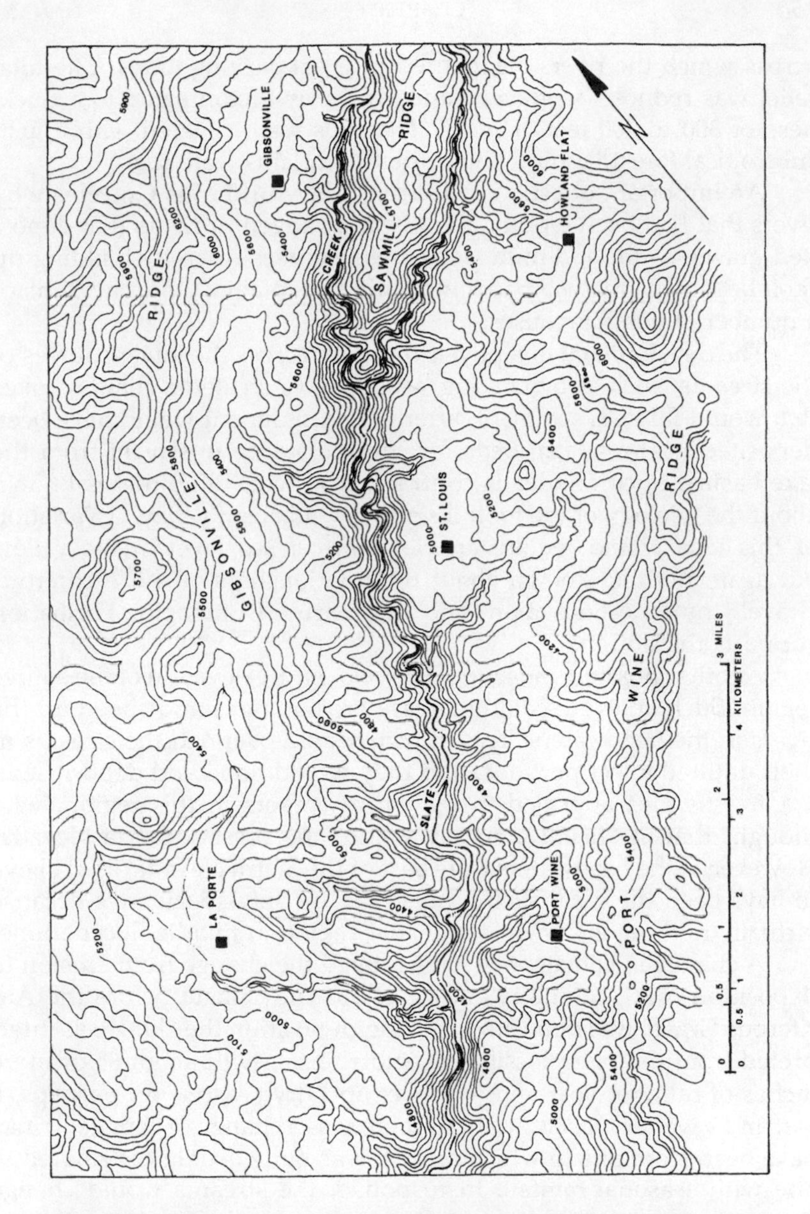

Fig. 86. The present landscape of the La Porte–Gibsonville District at the same scale and contour interval as fig. 85. The inner canyon of Slate Creek, 700 feet deep with side slopes as steep as 26 degrees, is typical of the modern canyons of the Sierra Nevada. Nothing like it is present in fig. 85. Only the gently rounded uppermost slopes of Gibsonville and Port Wine ridges cut in the easily eroded rock of the Penman Formation are comparable to the landscape of fig. 85.

across which the rivers meandered in great curving loops. The total relief was reduced as gravel, sand, and clay accumulated to a thickness of 300 to 400 feet or more. The hills were covered with a lush subtropical forest, and no doubt the floodplains were too.

An important question concerning the Auriferous Gravel is why rivers that had been eroding their beds changed to rivers that deposited gravel, sand, and mud along their courses, thereby building up their beds and greatly broadening their floodplains. One may think of a number of possible causes.

The development of a physical barrier across the lower courses of the streams, such as by faulting, would have created a chain of lakes that would fill with sediment, after which sediment would have been deposited in the stream beds for some distance upstream from the lake basins. Such an idea is contrary, however, to all that is known about the geology of the foothills and the adjacent valley. A variation of this idea would require an elevation of sea level in the valley, but again nothing known about the relationships of the Auriferous Gravel, the Ione Formation, and the marine Domengine Formation supports the notion.

Another possible cause could have been an elevation of the source region. Such an event could increase erosion rates and speed up the rivers in the headwaters. This in turn would overload the streams in their untilted lower portions, and they would flood and deposit sediment. This idea appealed to the early workers in this region, who thought that the headwaters lay within the present Sierra Nevada. However, if the headwaters were east of the Sierra Nevada, as I believe to have been the case, such a cause can be only speculative because virtually nothing is known about that region in middle Eocene time.

A change in climate may have caused the change from erosion to deposition, but nothing is known about the climate before the Auriferous Gravel was deposited. During deposition the climate as interpreted from the plant fossils was subtropical, with about 65 or more inches of rain per year and with wet and dry seasons or, perhaps, a wet and a wetter season. If, then, there was a change of climate, it may have been from one in which rainfall was quite evenly distributed to one with seasonal rainfall. In response, the streams would change from a regime of little fluctuation to one of periodic flooding with consequent deposition of sediment on floodplains. Between floods the streams would retreat to a normal channel upon a floodplain built up by successive episodes of flooding and deposition.

Such floods need not have occurred very often if they were the result of catastrophic precipitation such as happens, for example, in the central coastal region of Brazil. In the vicinity of Rio de Janeiro, a *tropical* but nearly subtropical region, catastrophic rains occur every few years. Rainfall is often as much as an inch per hour, and during some storms it has continued at that rate for many hours. Some of the resulting floods have stripped away the entire forest and soil from the underlying rocks and clogged the streams with debris, but the region affected by each flood is small. If such rainfall caused the Auriferous Gravel to be deposited, no one flood need cover a large area and floods need not have happened very often, but they would have needed to have affected every part of the region many times.

Since the period of deposition of the Auriferous Gravel may have lasted as long as 10 million years there was ample time for the requisite number of floods to have taken place. If one such flood occurred every 50 years there could have been 200,000 floods, or if floods occurred once every 20 years there could have been 500,000 floods. If each flood deposited an average thickness of only 2 feet of sediment over an area of 10 square miles, about 250,000 floods would bury the ten northern counties of the Sierra Nevada, an area of 12,500 square miles, to a depth of 400 feet. That seems to be within the realm of possibility, but of course all of this is mere speculation. The cause or causes of deposition of the Auriferous Gravel remains unknown. Perhaps it shall remain forever unknown.

Some final questions about the Auriferous Gravel concern the gold. Where did it come from? How much was recovered? How much remains? W. Lindgren, who first studied the Auriferous Gravel in great detail, concluded that the gold came from gold-bearing quartz veins in the metamorphic rocks east of, or upstream from, the places where it was found. He pointed out that such veins are common in the metamorphic rocks, a belt 30 to 40 miles wide east of the rich parts of the Auriferous Gravel, and that such veins are not common in the granitic rocks of the plutons that make up the higher, more easterly parts of the Sierra Nevada. He pointed also to the fact that Auriferous Gravel in the eastern parts that lay on the granitic rocks was not very rich in gold.

Lindgren believed that the gold was released from the veins by the extended period of weathering that preceded the deposition of the gravel. This is, of course, harmonious with the fact that most of the gold was in the deep gravel in the inner channel. Its accumulation there took place before most of the Auriferous Gravel was deposited. But there was some gold in bench gravel above the inner channel, and in a few places it was quite abundant. Since much of the bench gravel is now known to have come from east of the present Sierra Nevada it is possible that some gold also came from sources farther east.

The total amount of gold produced in California from placer deposits (not including that recovered by dredgers) in the period 1849–1923 is estimated to be about $1 billion, at the old price of $20 per troy ounce. That includes production from modern streams and production from the desert regions and northwestern California. Certainly, more than half of that came from the Auriferous Gravel, but owing to the absence of records no one will ever know how much the Auriferous Gravel did yield.

Some parts of the Auriferous Gravel contained little gold; other parts were quite rich. The area around La Porte is conceded to have been rich; it is reported that during the 16-year period from 1855 to 1871 not less than $60 million of gold at the old price was shipped from La Porte.

How much gold remains in the Auriferous Gravel? That is also unknown, but future miners, if there are any, are likely to be disappointed when they discover evidence of now unknown nineteenth-century underground workings by which the gold of the inner channels was removed.

LOVEJOY FORMATION

The Lovejoy Formation, next younger than the Auriferous Gravel, consists of a series of lava flows of black-appearing basalt whose most conspicuous occurrence is in the dark cliffs around Oroville Table Mountain (fig. 87). The same lavas in similar cliffs are present in a chain of occurrences that extend from Stony Ridge on the summit of the scarp overlooking Honey Lake to Oroville; all are part of a once continuous sheet, most of which has been removed by erosion. Thus the lavas extend across the Diamond Mountains, Grizzly Ridge, and the Sierra Nevada within a strip that was not less than 3 miles wide at Stony Ridge and perhaps as wide as 30 miles at the edge of the Sacramento Valley. The lava flows also underlie much of the Sacramento Valley south of Chico and are at the surface near Vacaville.

A good place to see the Lovejoy Formation is along the Beckwourth–Genesee Road for 3 miles northwest of its junction with the Bagley Pass Road in Sec. 5, T. 24 N., R. 13 E., on both sides of Red Clover Creek. There it is 400 feet thick and consists of at least nine lava flows that range in thickness from 10 to 50 feet (fig. 88). A close view of the lavas may be had at Oroville Table Mountain, where, by following the road from Oroville to Cherokee, one can drive across them and view the cliffs closely. At the west end of South Oroville Table Mountain the basalt is 300 feet thick and is composed of at least five and possibly eight flows (see fig. 87).

The Lovejoy lavas can also be seen very well on both sides of Grizzly Valley near the now inactive Walker Mine, along Lovejoy Creek from whence comes the name, and on Lumpkin and Mooreville ridges about 25 miles east of Oroville. The only place where the lava is directly on a main route is that on State Highway 70-89 just west of

Fig. 87. The Lovejoy Formation at the tip of South Oroville Table Mountain as seen from State Highway 70, at the Garden Drive overpass, Oroville. Here the Lovejoy Formation consists of at least five and possibly as many as eight lava flows with a total thickness of about 300 feet. The conspicuous black cliff, about 175 feet high, is made of the two uppermost flows. The rock of the slopes below the lava is the Ione Formation.

Fig. 88. The Lovejoy Formation. About nine flows of black basalt in a typical occurrence with cliffs and benches and scant vegetation. On the north side of Red Clover Creek, viewed from the Genesee–Beckwourth Road near its junction with the Mt. Ingalls Road. NE 1/4, Sec. 36, T. 25 N., R. 12 E.

Fig. 89. Lovejoy Formation. An unrepresentative occurrence where cliffs are not present. A road cut on State Highway 70-89 0.5 mile west of Lee Summit between Spring Garden and Cromberg.

Lee Summit between Cromberg and Spring Garden, where it shows as a bank of small fragments of purplish black rock (fig. 89).

The cliffs that result from the erosion of the lava flows are a characteristic feature. They are present at almost all known occurrences excepting only a few that are in heavy forest. For the most part the upper surface of the lava is a kind of local desert with few trees, scant grass, and some herbaceous plants. (An exception is Oroville Table Mountain, the top of which is a garden of wildflowers in spring.) The desertlike character is apparently owing to the fact that the lava is so closely fractured that rain and the meltwater of snow sink through it, leaving a surface that dries quickly. The chocolate-colored soil from the lava is thin and perhaps not very fertile.

The lava, which looks very dark from a distance, is on close examination dark brown, dark gray, or purplish on weathered surfaces. Freshly broken surfaces are dark gray or black. Vesicles are few, zones of broken and scoriaceous lava between flows are absent, and there are few large grains to be seen. The magma must have been very fluid. Columnar structure is common but the columns are not well defined. A fan of columns in lava along Red Clover Creek is shown in fig. 20.

The fractures or cracks in the lava are one of its most distinctive features. They are highly irregular curved surfaces coated with films of rusty iron oxide; they are not flat surfaces, and they do not occur in parallel sets as is the most common aspect of fracture surfaces in rock. The cracks are so pervasive that it is almost impossible to collect a piece as large as a baseball that will not easily break into smaller pieces along already-existing cracks. All of these features serve to distinguish the rock of the Lovejoy Formation from the basalt lavas of the younger Warner Formation.

The region occupied by the lava that now extends across several mountain ranges seems to have been a broad, shallow valley. One or more streams flowed there, for in many places gravel is present beneath the lava. Probably the landscape was one of low relief like that upon which the Auriferous Gravel was deposited. Streams also flowed over lava, because in some places a little sand and gravel are present between lava flows. Although there is no direct evidence, there can be little doubt that the lava flowed from northeast to southwest.

The gravel below the lava is darker colored than that of the Auriferous Gravel and it contains less quartz. Some of it contained enough gold to interest miners but, judging by the absence of large workings, not enough to make profitable mines. Because the lava also rests in places on the Auriferous Gravel, it is possible that the gold in the younger gravel was pirated from the older formation by erosion.

Apart from the gravel at the base, which is to be included as part of the Lovejoy Formation, the lava rests on various granitic and metamorphic rocks, on the Auriferous Gravel, and on the Ione Formation. For that reason it is younger than middle Eocene.

The youngest possible age of the Lovejoy is fixed by the fact that two very large masses of Lovejoy basalt on the floor of the hydraulic pit of the Upper Dutch Diggings in Auriferous Gravel at La Porte were exhumed by the mining (figs. 83 and 90) and numerous smaller pieces are embedded in the sandy clay of the pit wall. These were emplaced in the sedimentary rock by erosion and deposition before the leaf-bearing volcanic ash present at the top of the wall of the pit was deposited. The ash, previously mentioned in connection with the Auriferous Gravel, has been dated radiometrically as about 29 million years old, well before the end of the Oligocene epoch. Accordingly the Lovejoy Formation must be older than that, perhaps early Oligocene—30 to 40 million years old.

The extent of the Lovejoy Formation is now quite well known, but the source of lava has never been found. Perhaps it is concealed beneath the younger rocks of Honey Lake Valley, or perhaps it was still farther east or north. Those regions are also extensively covered by younger rock so perhaps the source will never be discovered. In any event the lavas of the Lovejoy Formation are quite remarkable for it is apparent that they flowed at least as far as from Stony Ridge to

Fig. 90. A block of basalt from the Lovejoy Formation, about 12 feet in diameter, formerly embedded in the Auriferous Gravel at the Upper Dutch Diggings, La Porte, now near the base of the cliff shown in fig. 83.

Vacaville, an airline distance close to 150 miles. They covered an area of more than 50,000 square miles to a depth of as much as 600 feet. If the average thickness is as much as 400 feet, a likely value, the volume of lava erupted was not less than 37 cubic miles.

Excepting the occurrence of cobbles of unmetamorphosed volcanic rock in the Ione Formation at Oroville Table Mountain, and a few also in the Auriferous Gravel, all of unknown source, the lavas of the Lovejoy Formation are the earliest volcanic rocks in the Sierra Nevada erupted after the Nevadan Orogeny. They are the first of a long-continued sequence of eruption and deposition of rock of volcanic origin.

INGALLS FORMATION

The Ingalls Formation, named for its occurrence on the northeast slope of Mt. Ingalls, consists of dark gray volcanic mudflow breccia that is almost devoid of obvious layering except at the base, where, in one place, conglomerate about 30 feet thick is exposed to view (fig. 91). The conglomerate is composed of boulders of volcanic rock and

Fig. 91. Ingalls Formation. Conglomerate below and mudflow breccia above with a discontinuous layer of sand between. Near the base of the formation. See fig. 93 for location.

some of granitic rock like that presently visible in place nearby. The angular fragments of the breccia, many as large as 2 feet in diameter, are mostly dark gray andesite. Some blocks are basalt, and a few of these were eroded from the Lovejoy Formation. Because mudflow breccia is dominant also in the younger Bonta and Penman formations, and because intrusions of volcanic breccia are abundant in this region, the discussion of their origin follows that of the descriptions of the younger formations. However, no source for the volcanic material of the Ingalls Formation has been identified.

The best place to see the Ingalls Formation is in lower Clover Valley along the Beckwourth-Genesee Road northwest of its junction with the Bagley Pass Road in Sec. 5, T. 24 N., R. 13 E. For about 2 miles along that road the black, craggy, weirdly shaped outcrops of the Ingalls, as shown in fig. 92, stand conspicuously above a very thin and poor stony soil. From near the junction of the Beckwourth-Genesee Road with the road to the summit of Mt. Ingalls, one can see on the northeast side of Red Clover Creek the excellent exposures of the Lovejoy Formation, on top of which is the Ingalls Formation with gravel at its base. Both formations are horizontal (fig. 93). The Ingalls

Fig. 92. Ingalls Formation. Rough, craggy black outcrops of mudflow breccia typical of the Ingalls Formation. Beside the Beckwourth–Genesee Road along Red Clover Creek in the SW 1/4, Sec. 32, T. 25 N., R. 13 E.

is about 550 feet thick; its top is not visible from this point, but about a mile to the east it is overlain by younger formations.

The Ingalls is also easily visited on the south side of Grizzly Ridge along the road from Long Valley to Happy Valley where it passes along the east side of the canyon of Little Long Valley Creek in Sec. 1, T. 23 N., R. 11 E. There it rests on Auriferous Gravel (fig. 94). Near its bottom it contains blocks of basalt from the Lovejoy Formation as large as 12 feet in diameter. One concludes from these facts that the Lovejoy Formation at that point was removed by erosion before the Ingalls was deposited on the Auriferous Gravel, but that some Lovejoy from a place not far away was incorporated into breccia of the Ingalls.

Such large blocks of basalt of the Lovejoy Formation must be thought of as being quite fragile because of the close spacing of cracks. One may suppose that the blocks were removed from a cliff, not necessarily a very tall one, either along a stream course or one produced by faulting. The Ingalls does rest on an erosion surface, but as far as is known that surface shows very little irregularity—none that could be called either a stream-cut valley or a cliff caused by faulting. It is true that no faults younger than the Nevadan Orogeny have been found

Fig. 93. Ingalls Formation. The north side of Red Clover Creek in Sec. 30, T. 25 N., R. 13 E. Telephoto view from the Mt. Ingalls Road. The upper cliff is the mudflow breccia of the Ingalls Formation. The lower cliffs are in lava flows of the Lovejoy Formation. The distinct horizontal line at the base of the upper cliff is the contact between conglomerate and mudflow breccia shown in fig. 91. The bottom of the conglomerate, which is the unconformity between the Ingalls and Lovejoy formations, is at the base of the tree at the center of the picture, indicated by the black arrow.

that pass under the Ingalls Formation and do not cut it. However, considering the relationship of the Ingalls rocks in the region between Red Clover Creek and Little Long Valley Creek, there can be no doubt that faulting did occur, as is shown diagrammatically in fig. 95. Of course, no single fault is responsible for the relationship—there are many faults between Red Clover Creek and Little Long Valley.

This episode of faulting terminated the long period of apparent tranquility of the earth's crust that began at the end of the Nevadan Orogeny and lasted through the Cretaceous period, the Paleocene epoch, and much of the Eocene epoch, a span of at least 40 million years. At the present time, this is the earliest known evidence of a period of renewed crustal unrest that culminated millions of years later in the Sierra Nevada and the nearby mountains as they are known today.

Fig. 94. Ingalls Formation on the ridge west of Little Long Valley Creek. Here the Ingalls rests on the Auriferous Gravel that is visible in the lower left corner of the picture. NE 1/4, Sec. 1, T. 23 N., R. 11 E. By the forest road from State Highway 70-89 at Rattlesnake Creek to Happy Valley.

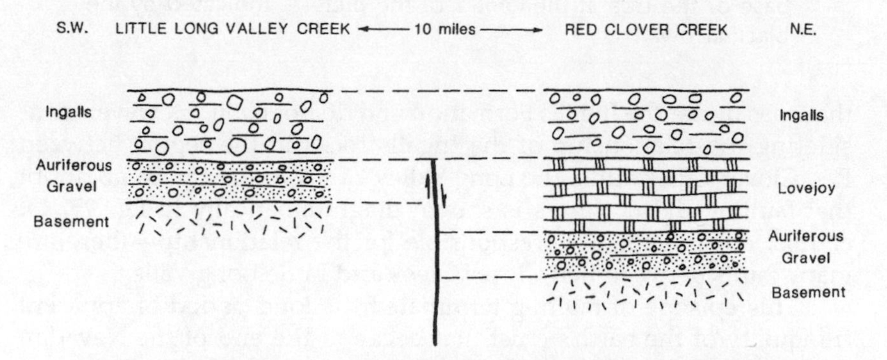

Fig. 95. The relationship of the Ingalls Formation to older formations on the ridge west of Little Long Valley Creek and at Red Clover Creek. Because the formations are horizontal and because the Lovejoy was once a continuous sheet, its absence west of Little Long Valley Creek can be explained only by faulting and erosion before the deposition of the Ingalls.

The presently known extent of the Ingalls Formation is not large. From its occurrence along Red Clover Creek in the Diamond Mountains it extends but a little distance both northwest and southeast; the region to the north has not yet been studied. The Ingalls has not been found in the Sierra Nevada, but it is present as far south as Sloat, which is virtually at the east base of the Sierra Nevada.

The limited extent of the Ingalls may be accounted for by the fact that its deposition was followed by several periods of faulting and erosion. Because no fossils have been found in it and no attempt has been made to determine its age by radiometric methods, its age is not closely determined. It is younger than the Lovejoy Formation, which is late Eocene or early Oligocene, and it is older than the overlying Delleker Formation, which is late Oligocene and early Miocene. Possibly it is Oligocene.

The only other rock unit to which it might be related is the Wheatland Formation, present northeast of the town of Wheatland at the edge of the Sacramento Valley (see figs. 79 and 80). The Wheatland Formation consists of conglomerate and sandstone, with pebbles of basalt and of andesite similar to the andesite of the Ingalls, and it contains marine fossils of Oligocene age. This tenuous connection lends some support for an Oligocene age for the Ingalls, since both formations are the result of andesitic volcanism before the eruption of the rhyolitic rocks of the Delleker Formation. There is at least a possibility that the Ingalls and Wheatland formations were once continuous. If that were so, the Ingalls was either eroded from the region in between or it has not yet been recognized there.

DELLEKER FORMATION

The widespread Delleker Formation consists mostly of silicic volcanic ash, equivalent to the igneous rock rhyolite, that has been compacted into firm rock. Some gravel is present at the base.

The compacted ash or tuff, as it is commonly known, is variously colored—white, light gray, cream, less commonly pink—and much of it is stained brown by iron oxide. It yields a thin and stony soil of poor quality. Although the appearance is far from uniform, it is everywhere characterized by sparkling clear crystals of quartz and shiny hexagonal flakes of the black mica biotite. It is unstratified.

Most of the rock of the Delleker is a hard, lusterless, stony-appearing rock that is intricately cracked in a pattern that reminds one of the surface of crazed pottery (fig. 96). The cracks, however, penetrate the entire rock. This type of rock can be easily reached along the Mohawk–Chapman logging road about 3.5 miles from its junction with the Gold Lake Highway at the Frazier Creek bridge, and also

Fig. 96. Delleker Formation. The most common aspect of the welded tuff is this crazed or crackled appearance. A close view of the rock in fig. 4. Along the Mohawk–Chapman logging road, NW 1/4, Sec. 2, T. 21 N., R. 12 E.

across the road from the now-collapsing log cabins by the road from Gold Lake to Mills Peak. The first location is in Sec. 2 and second in Sec. 10, T. 21 N., R. 12 E. At both places, and elsewhere where this rock occurs, it contains irregularly shaped cavities about 0.5 inch in diameter whose surfaces are striated. I believe them to be the molds of pieces of charcoal; by breaking a sufficient amount of rock one can find some that contain carbon.

Another common type of rock consists of pieces of cream-colored pumice that are flattened as a result of the collapse of their tubes, embedded in a light gray matrix that at first glance looks sandy (figs. 97 and 98). On close inspection of freshly broken surfaces, the matrix is seen to consist of wispy-shaped fragments of broken pumice that have been welded together so that the entire broken surface has a glassy luster. This kind of rock is easily visited on barren rock surfaces that are across a shallow ravine on the north side of the Willow Creek Road, near the center of Sec. 32, T. 23 N., R. 13 E., about 1.75 miles from its junction with State Highway 70 (fig. 99).

An uncommon variety that occurs near Haskell Peak superficially resembles obsidian—it is compact black glass. Only under the microscrope is it revealed to have been ash.

Fig. 97. Delleker Formation. Gray, granular-appearing welded tuff with flattened, cream-colored, once angular fragments of pumice shown here on edge. For location see fig. 99.

All of these rocks are quite clearly members of an interesting group called welded tuff or pyroclastic flows. I prefer the former term because whereas pyroclastic flow expresses how the material got where it is, welded expresses what happened after it got there—how it became a hard rock. Moreover, not all pyroclastic flows are welded. Welded tuff is also known by the unlovely name of ignimbrite.

The eruption of pyroclastic flows has been observed many times. They were, for example, the principal product of the eruptions in 1980 of Mt. St. Helens, Washington. A pyroclastic flow consists of a mass of pieces of pumice of all sizes from perhaps a foot in diameter down to the finest powder and gas—that is, water vapor—so hot that the mixture may be incandescent. Such mixtures are the result of vio-

Fig. 98. Delleker Formation. The same gray, granular-appearing welded tuff shown in fig. 97. The light-colored piece of pumice at the center is the flattened surface of a once angular block. The original silky appearance of the pumice remains although the tubular vesicles are collapsed. The very black material is a coating of lichen. For location see fig. 99.

lent explosions. They are turbulent and they flow across the land not like a cloud of windblown dust, but more like a flood of water. The flow may come either directly from the side or base of a plug dome or it may be the material that falls back from a mass blasted directly upward from the volcanic vent. Flows travel at speeds estimated to be as high as 100 miles per hour. They blanket the land and may even travel upslope and across ridges.

No observed pyroclastic flow has become welded but most volcanologists believe that the welded tuffs did originate from these flows. Perhaps the problem is one of size, that welded tuffs originate only from pyroclastic flows much larger than any yet observed.

It is the residual heat in the deposited tuff of fragmented material that permits the pumice blocks to collapse into slabs, allows the finer particles not only to stick together but also to deform, and, in extreme cases, causes some of the welded tuff to resemble obsidian.

A further process often takes place in welded tuffs. After welding, the glass devitrifies; that is, it crystallizes to an extremely fine grained aggregate. That is what causes the dull, stony, or lusterless

Fig. 99. Delleker Formation. The excellent and interesting exposure of the gray, granular-appearing welded tuff shown in figs. 97 and 98. Lack of soil and poor fertility have resulted in stunted trees. On the north side of the Willow Creek forest road near the center of Sec. 32, T. 23 N., R. 13 E. The figure of a man at the center provides a scale.

appearance of so much of the rock of the Delleker Formation, and of welded tuff elsewhere.

Features of welded tuffs other than those described are present in the much thicker rocks of the same composition and geologic age in Red Rock Canyon, Nevada, east of U.S. Highway 395 about 6 miles north of Hallelujah Junction, and in the ranges west of Pyramid Lake. These and similar rocks near Virginia City, Nevada, are called the Hartford Hill Rhyolite.

The thickness of a single pyroclastic flow deposit can be as little as a few inches or as much as hundreds of feet. The real extent of single flows is not well known, but thick sequences of welded tuffs cover areas larger than 10,000 square miles. Their distance of travel is also uncertain. Some recent ones are known to have traveled as far as 35 miles, and some slightly older are thought to have traveled 60 miles. No information about their dimensions has been obtained from the Delleker Formation.

Although no possible sources for the Delleker have been found in eastern Plumas County, intrusions of rhyolite near Pyramid Lake could have been the source, and also the source for the Hartford Hill Rhyolite.

Welded tuff is present near Alta just off Highway I-80. If this tuff came from the vicinity of Pyramid Lake, it would have had to travel 80 to 90 miles. Welded tuff is also present near Mokelumne Hill in the central Sierra Nevada, in the Valley Springs Formation, which is the same age and composition as the Delleker. If that came from Nevada, which is by no means certain, it would have traveled about 80 miles. These possible distances of travel do not seem excessive.

Water-laid rhyolite ash of the same age as the Delleker is interbedded with welded tuff in the western Sierra Nevada and as far west under the Sacramento Valley as the Sutter Buttes. The Delleker and Valley Springs formations and the Hartford Hill Rhyolite are part of a great sheet composed of many flows of rhyolite ash that spread across the site of the Sierra Nevada, traveling westward from sources in or near to the state of Nevada. It rests on an erosion surface—an unconformity.

Streams flowed on subaerial surfaces of unconformity such as that at the base of the Delleker Formation, so gravel is to be expected there. At most places the amount of gravel is small—only scattered cobbles and pebbles. Pebbles in the ash in some places indicate that fast-moving ash flows were capable of picking up rock from the surface over which they traveled.

Abundant cobbles and pebbles of quartzite, a metamorphosed quartz-rich sandstone, occur in the Delleker in the hills west of Portola and north of State Highway 70. The underlying basement here is the granitic rock of the eastern plutons, which contains only small areas of metavolcanic rocks; none of quartzite are known. The quartzite probably came from east of the plutons, perhaps from east of Pyramid Lake.

On the east side of the crest of the Sierra Nevada between Mills Peak and Haskell Peak, the Delleker welded tuff is quite thick, and so also is the gravel—as much as 200 feet. Gravel is visible in the road cuts of the Mohawk–Chapman logging road and its branches that lead to Chapman Saddle. The same gravel is also present, but not so thick, on the south side of the crest in Sec. 15 along the road from Gold Lake to Mills Peak. The gravel contains a wide variety of metamorphosed volcanic rocks, some quartzite, and boulders of granite. The granite (used here in the strict sense of a potassium-rich granitic rock) is unlike the granitic rock on which the gravel rests; the nearest known occurrence of similar rock is north of Sierra Valley and east of there in Nevada.

Thus, both tuff and gravel provide evidence for transport of material westward or southwestward toward and across the site of the

Sierra Nevada. There was as yet no mountain range, although, as indicated earlier, the Sierra Nevada may have been outlined by faulting before the Ingalls Formation was deposited.

Possibly, however, the lower part of this thick gravel near Haskell Peak is Auriferous Gravel; perhaps only the upper part, in which gravel and ash are mixed, belongs to the Delleker Formation. The lower gravel does contain gold, and it has been mined on a small scale on both sides of the crest, but nowhere is it rich in quartz. In any case, the evidence for transport to the southwest is clear, and since this is some of the gravel once believed to mark the course of the Jura River of middle Eocene age it is important to note clearly that the stream that deposited the gravel did not flow to the northwest.

The age of the Delleker Formation, of the Valley Springs Formation of the central Sierra Nevada, and of the Hartford Hill Rhyolite of western Nevada is well established by the radiometric dating of minerals in the tuff by the potassium-argon method. Nine determinations fall in the range of 16 to 25 million years, which is early Miocene, and five determinations are in the span of 25 to 33 million years which is late Oligocene. Evidently the period of eruption of rhyolite also spanned the boundary between the Oligocene and Miocene epochs.

The deposition of the Delleker Formation in eastern Plumas County was followed by a period of faulting. This was succeeded in turn by erosion that affected most, if not all, of the northern Sierra Nevada and Diamond Mountains. Only remnants of the Delleker were preserved, mostly in the lower-standing fault blocks, and all the older formations were once again exposed at the erosion surface. Of course, the older formations were also eroded at this time. This is the surface upon which the succeeding Bonta Formation was deposited.

Weathering of the newly exposed rocks penetrated deeply below the erosion surface, affecting especially the granitic rocks, which became a mass of partly decayed loosened granules like that shown in fig. 5 to a depth in some places of 200 feet. This had an important consequence for the basalt lavas of the Warner Formation, described further on.

The relief on the erosion surface is so low that almost none can be detected, but in only a few places is the exposure of rock adequate for detailed knowledge. One such place is along the railway track between the Willow Creek trestle and tunnel 36, and there the erosion surface has only about 20 feet of relief.

BONTA FORMATION

The Bonta Formation rests on the remarkably smooth erosion surface cut across the Delleker and older formations (fig. 100). It was

Fig. 100. Bonta Formation. The unconformity at the base of the Bonta Formation. Andesite mudflow breccia about 10 feet thick rests across the ends of deeply weathered metamorphic rock. Along the railroad, east of Willow Creek. The camera was pointed upward. NW 1/4, Sec. 19, T. 22 N., R. 13 E.

deposited as a widespread sequence of beds, mostly of gray volcanic mudflow breccia and volcanic conglomerate, that had incorporated in them angular fragments, pebbles, cobbles, and boulders of older rocks.

The mudflow breccia (fig. 101) is a compact rock that consists of angular and slightly rounded blocks of andesite, many of which are 2 to 3 feet in diameter, with a few as large as 10 feet, between which are successively smaller pieces down to sand- and mud-size particles. Each mudflow layer is a single depositional unit within which there is no internal sorting or stratification. Most mudflow layers contain a large variety of andesites of somewhat different appearance, but some are composed of pieces that are all alike. Most of the andesite contains phenocrysts of a shiny blackish brown variety of the mineral hornblende.

Beds of volcanic conglomerate are interlayered with the mudflow breccia. Most of the rocks in them are andesite like those of the mudflow breccia, but of course they are more rounded—some very well rounded, but others only poorly so. Most of the beds of conglomerate also contain sand and mud so that they too are compact rocks (fig. 102). Both mudflow breccia and conglomerate contain fossil wood.

Fig. 101. Typical mudflow breccia of the Bonta Formation. The lower bed has smaller blocks and more abundant matrix than does the upper bed. Note lack of rounding, sorting, and internal stratification. The figure in the lower center serves as a scale. Along the railroad, NE 1/4, Sec. 19, T. 22 N., R. 13 E.

Fig. 102. Bonta Formation. Volcanic conglomerate resting on mudflow breccia. The thickness of the gravel is about 20 feet. Below the bush the long dimension of the cobbles is inclined downward to the right. This indicates that the depositing current moved toward the left, which is southwest. Along the railroad east of Willow Creek, NW 1/4, Sec. 20, T. 22 N., R. 13 E.

Fig. 103. Bonta Formation. A block of granitic rock estimated to be 15 by 25 by 32 feet in mudflow breccia. Note seated figure for scale. Below the Jackson Creek Road about 0.25 mile northeast of the viewpoint marked on fig. 156. NW 1/4, Sec. 20, T. 23 N., R. 12 E.

Fragments of older rocks, mostly in the mudflow breccia, but also in the conglomerate, are recognizable as having been eroded from the Lovejoy, Ingalls, and Delleker formations, and, whereas fragments of metamorphic rock are not very common, those of granitic rock are exceedingly abundant.

Rounded boulders of granitic rock as much as 5 feet in diameter are present nearly everywhere in the lower part of the formation. Some are excessively large; one that measures about 15 by 25 by 32 feet, as large as a modest-size house, and that must weigh more than 1,000 tons, is shown in fig. 103. This is a giant block, but it is not surprising that such a piece could be moved, because about 75 percent of the weight of any piece of granitic rock would be supported by the bouyancy of the mudflow. Finer debris from granitic rock is so abundant that flakes of mica and grains of quartz are present everywhere in the lowest several hundred feet of the Bonta.

Blocks of tuff from the Delleker Formation as much as 50 feet thick and 100 feet long were incorporated into the Bonta. Some of that size are present on the south slope of Penman Peak north of Blairsden. The abundance of smaller pieces of tuff from the Delleker is indicated by fig. 104. Well-rounded boulders, as large as 2 feet in diame-

Fig. 104. Bonta Formation. Andesite mudflow breccia that contains many blocks of white rhyolite welded tuff reworked from the older Delleker Formation. A man stands at the center of the picture for scale. Along the abandoned highway above State Highway 70, 1 mile east of the junction of State Highways 70 and 89 near Blairsden. NW 1/4, Sec. 14, T. 22 N., R. 12 E.

ter, of welded rhyolite tuff unlike any known to be present nearby, but which are like those in Red Rock Canyon and in the mountains west of Pyramid Lake, are especially abundant in the W 1/2 Sec. 28, T. 23 N., R. 12 E., on the north side of the canyon of Consignee Creek. Possibly they were transported westward a distance not less than 35 miles (fig. 105).

Layers of sand composed of bits of andesite and mineral grains of the andesite are present here and there in considerable amounts although the total is minute compared to the vastly greater amounts of mudflow breccia and conglomerate. Chocolate-colored volcanic mud in layers as thin as a half inch or as thick as several feet also occur.

Lava flows of hornblende andesite are present in the Diamond Mountains north of Sierra Valley, but there are none on Grizzly Ridge.

The soil over the Bonta Formation is thin on steep south slopes such as that below Penman Peak, but where there is a soil it is rust colored with a distinct crust that forms when the soil dries in the spring or after a summer rain.

The sum of these features gives the Bonta Formation a distinctive character, but it is true that in places where only a little of it can be examined it could be mistaken for either the older Ingalls Formation

Fig. 105. Bonta Formation. Well-rounded boulders of welded tuff derived from the older Delleker Formation or its equivalent in Nevada, the Hartford Hill Rhyolite. Rock like that of these boulders is not known in place west of the California-Nevada boundary. The hammer handle is 13 inches long. One mile east of Two Rivers on a forest road above Consignee Creek. W 1/2, Sec. 28, T. 23 N., R. 12 E.

or the younger Penman Formation, both of which are also mostly andesite mudflow breccia.

The Bonta Formation can be seen with its typical features along paved roads at three places, one of which is along County Road A-15, which joins State Highway 89 about 1.5 miles southeast of Clio. In the road cuts about 1 mile up A-15 from the junction, typical Bonta mudflow breccia contains very large blocks of granitic rock, of tuff from the Delleker Formation, and of andesite mudflow breccia similar to that which encloses it. Another is the first road cut on State Highway 70 west of Portola, where compact volcanic conglomerate contains an abundance of large pieces of granitic rock and tuff from the Delleker Formation. A third place, well worth a visit for the scenery alone, is along the road from Chilcoot to Frenchman Lake. The road follows Little Last Chance Creek through a narrow box canyon in the Bonta Formation, which is exposed in spectacular cliffs.

Other places to see the Bonta are along unpaved but maintained forest roads that may be rough and rocky at times. One can see very large blocks of hornblende andesite such as the one shown in fig. 106,

Fig. 106. Bonta Formation. An unusually large block of hornblende andesite in mudflow breccia. It measures 12 by 9 by 5 feet. About 1,000 feet southwest of the intersection of the forest road from Lake Davis to Mt. Ingalls with an abandoned road along the ridge north of Lake Davis. SE Cor., Sec. 12, T. 24 N., R. 12 E.

small blocks of pink granite (see fig. 9), mudflow breccia of a single variety of andesite, and layers of cross-bedded sand near the road junction at the SE Cor., Sec. 12, T. 24 N., R. 12 E. This location, on the crest of Turner Ridge,[2] is reached by following the road that leads to the Mt. Ingalls lookout for about 2.5 miles from its junction with the Walker Mine Road on the north shore of Lake Davis. Here also the Penman Formation, next younger than the Bonta, rests on the Bonta, and the two can be compared as is related in the description of the Penman Formation. The Bonta Formation on the southwest slope of Penman Peak can be reached from a road that crosses the mountain front at about 5,000 feet elevation. The road is the continuation of the

2. Turner Ridge extends from Sierra Valley at Beckwourth to Genesee Valley. It is northeast of and roughly parallel to Grizzly Ridge. Crocker Mountain is at its southeast end and Mt. Ingalls at its northwest end. The name does not yet appear on official maps of the U.S. government, but it was approved by the U.S. Board on Geographic Names on February 28, 1975. The name is in honor of Henry Ward Turner (1857–1937), who as geologist with the U.S. Geological Survey did the initial geological studies in this area.

Denton Creek Road at its southeast end and a branch of the Bonta Ridge Road at the northwest end.

An excellent viewpoint of landscape in general, and of the Bonta Formation in particular, in the NW 1/4 C, Sec. 20, T. 23 N., R. 12 E., can be reached by following the road from the Jackson Creek campground to Happy Valley and taking the first branch to the right (to the southeast), which is at the top of the grade. About a half mile from the intersection the road leaves the forest and comes out on a barren slope. One should stop at a sharp left turn, facing Mt. Jackson. Here typical Bonta conglomerate is exposed in the road cut and in broken blocks at the outer edge of the road. To the west one looks down the canyon of the Middle Fork of the Feather River; to the south, across Big Hill to Eureka Ridge, which is the crest of the Sierra Nevada; and to the southeast, across Jackson Creek to Mt. Jackson, capped with lava flows of the Warner Formation. Below and to the right of Mt. Jackson is a black crag that is a plug of basalt, one of the many feeders for the lavas of the Warner Formation. The slope of Mt. Jackson below the cap of lava flows is stratified volcanic conglomerate and conglomeratic mudflow breccia of the Bonta Formation (fig. 107). Large-scale stratification such as this is common everywhere in the Bonta but in few places is it so well displayed; it is best seen from a distance. Both above and below the next half mile of this road there are many large, rounded pieces of granitic rock, including the giant block shown in fig. 103. It is below the road and a small tree grows on it.

Well-bedded Bonta like that on Mt. Jackson is also present in the canyon of Little Last Chance Creek and on the west face of Dixie Mountain.

The Bonta Formation is widespread in eastern Plumas County. It is present over most of Grizzly Ridge southeast of Argentine Rock, where it is 1,000 feet thick. It occurs on the lower slopes of Mt. Ingalls and around Clover Valley. It makes up most of Turner Ridge between Mt. Ingalls and Crocker Mountain. It is widespread in the mountains north of Sierra Valley east to the Honey Lake scarp, and at Dixie Mountain in that region it is 2,000 feet thick. It also occupies the lower slopes around Beckwourth Peak. Rocks of the same age and similar kind are present over much of northwestern Nevada, where they are also 2,000 feet or more thick.

There is some Bonta in the Plumas Trench. It is present on the hill occupied by the Calpine lookout and in Long Valley. However it is absent from Big Hill, Spring Garden, and American Valley.

Only a little Bonta has been recognized on the Sierra Nevada. A small area of it, less than 100 feet thick, is present on the ridge southeast of the Mohawk Saddle, in Sec. 13, T. 21 N., R. 12 E.; a second and

Fig. 107. Bonta Formation. Part of the northwest slope of Mt. Jackson as seen from the Jackson Creek Road at the viewpoint marked on fig. 156. Large-scale bedding such as this is easily recognized at a distance but is almost invisible close up. View from NW 1/4, Sec. 20, T. 23 N., R. 12 E.

more important occurrence, because it contains fossil leaves, is on the ridge between the Lakes Basin campground and Frazier Creek.

The Miocene age of the Bonta Formation was first established in 1891 by the study of fossil leaves found in mudstone along the Middle Fork of the Feather River near Clio, although at that time the mudstone was mistakenly believed to be a part of the Mohawk Lake Beds, which are now known to be very much younger. For many years the location of the fossil leaves was unknown but it was rediscovered in the 1950s, a new collection was made, and it was then recognized that it was in the very lowest part of the Bonta Formation. Unfortunately, a few years later, probably in 1963, high water on the river completely removed all remaining parts of the leaf-bearing mudstone. However, in the 1960s, when the Gold Lake Highway was constructed, rock containing the same assemblage of leaves was encountered in one of the road cuts near Gold Lake. Both collections indicate a middle Miocene age of the lowermost part of the Bonta Formation. Still later a radiometric date of 19 million years was obtained from a carefully selected sample of andesite from a mudflow breccia. This is also near the middle of the Miocene and confirms the ages based on fossil leaves.

A number of assemblages of fossil leaves found in western Nevada are in rock similar to, and the same age as, the Bonta Formation. They help us understand the character of the country at that time and the conditions under which the Bonta Formation was deposited.

According to D. I. Axelrod, the leaves from Clio and Gold Lake indicate that the region was not more than 2,000 feet above sea level when they were deposited. In conjunction with fossil leaves from farther east they indicate that the climate was rather mild, with an annual rainfall of 35 to 40 inches. That there was a summer rainy season as well as winter rains is indicated by the presence of trees such as liquidambar and magnolia that are no longer native to the West, but are native to southeastern United States, where it does rain in the summer.

An elevation of 2,000 feet for a place that is now the summit region of the Sierra Nevada tells us that there was no Sierra Nevada at that time. The fossil leaves from western Nevada, according to Axelrod, indicate the absence of a pronounced rain shadow—that is, a strip of land of reduced rainfall on the lee side of a mountainous area such as exists now just east of the Sierra Nevada.

With these facts in mind it is possible to give an impression of the appearance of northeastern California and northwestern Nevada during the time that the Bonta Formation was deposited. At the beginning, there was the remarkably smooth erosion surface already mentioned, across which westward-flowing streams drained into the Sacramento Valley; there was no barrier to westward flow of water or transport of sediment and no barrier to the eastward course of storms.

Volcanic eruptions occurred from many vents that are now represented by intrusions, some of which are of massive andesite and many of which are of andesite breccia. Although some of the known vents were sources for the younger Penman Formation, it is not yet possible in all cases to distinguish them from those that supplied the Bonta Formation. The matter is discussed further on.

Soon after deposition began, the fossil leaves found at Gold Lake and Clio were deposited in what were no doubt ponds on the floodplains of rivers. Elsewhere, leaves were entrapped in similar clay or in volcanic ash or diatomite in lakes at many places in western Nevada and interior Oregon. Volcanic eruptions produced lava, ash, and volcanic breccia that became cobbles, pebbles, sand, and mud that were transported by streams or mudflows and deposited in the forms already described. Stream courses also provided the channels down which mudflows traveled. This condition prevailed until a thickness of more than 2,000 feet was added to the land. Now it must not be thought that this would add 2,000 feet to the altitude of the land because the process of isostasy described in Chapter 3 operates all the time. Accordingly the net increase in elevation was probably no more than 200 or 300 feet. The volcanoes did not constitute a mountain chain, for evidence drawn from plant fossils found farther east indicates the absence of a rain shadow and therefore precludes a volcanic barrier as well as an uplifted Sierra Nevada. Perhaps there were volcanoes as much as 1,000 feet high; certainly there were no giant cones like Mt. Shasta.

The fossil plants show that there were forests marginal to rivers and lakes, open woodlands, and forests on the higher areas composed of deciduous hardwoods and coniferous trees. Among the latter was the giant Sierran redwood, then widespread, but now restricted to a few small areas on the western slope of the Sierra Nevada, where it is often referred to as a "living fossil." Five logs (not petrified), one 50 feet long, and much vegetative material of the Sierran redwood were unearthed during the excavations in preparation for the dam at Frenchman Reservoir. Farther north the Coast redwood grew inland across Oregon and Idaho.

Although the bones of animals have not yet been found in the Bonta Formation, they have been found in rocks of about the same age farther south in the Sierra Nevada, in the Coast Ranges of central California, and in Nevada. Included are the ancestors of horses and camels that, like the modern forms, ate grass. Browsers, such as deer, and aquatic beavers are represented among a host of other animals, large and small. The animals, like the plants, indicate a variety of habitats and vegetative types over a very large region.

How long a time passed during the deposition of the Bonta Formation and its related rocks cannot be said with any precision but it was probably on the order of 5 to 10 million years.

Volcanic activity continued throughout the period of deposition of the Bonta and for long after, but it should be understood that at no time was the country a barren wasteland. One may well ask how a thickness of 2,000 feet of volcanic rock can be added to a large tract of land that is well watered, covered with vegetation, and populated by a large assemblage of animals both large and small without devastation. The answer to the question is, of course, that devastation did occur. Shallow valleys became filled with lava, mudflows, volcanic conglomerate, and other volcanic debris. Streams were repeatedly diverted to new channels, and lakes were created and obliterated. The land was blanketed from time to time with layers of volcanic ash from violent explosive activity. But not all the land was devastated at any one time, and geologic time is long. Two thousand feet in 5 to 10 million years is an average of only five- to ten-thousandths of a foot per year. It should not be surprising that infrequent short-term local catastrophes should integrate into long-term widespread tranquility. After all, the processes just described are going on today in regions as diverse as the tropical jungle-covered islands of Indonesia, and the near-arctic forested Alaska Peninsula. Both are well populated with animals, and in both areas both plants and animals survive intermittent volcanic catastrophes.

One important question concerning the former extent of the Bonta Formation remains: did it or did it not extend across the present site of the northern Sierra Nevada? There is much hornblende andesite mudflow breccia and conglomerate on the western slope in the basins of the Feather and Yuba rivers, but it is like the younger Penman Formation; it lacks the characteristic features of the Bonta. It seems that if the Bonta Formation was deposited across the site of the northern Sierra Nevada, almost all of it was removed by erosion before the Penman Formation was deposited. There are several reasons for believing that this is what happened.

There *is* some Bonta on the Sierra Nevada near Gold Lake and at the Mohawk Saddle. The rock materials of the mudflows and volcanic conglomerate of the Bonta *did* travel westward; there was no barrier to their transportation across, and deposition on, what is now the Sierra Nevada. The relationship of the Bonta Formation in the Plumas Trench and on Grizzly Ridge aid in understanding the history. A thickness of a few hundred feet of Bonta Formation is present on the Plumas Trench in Long Valley, and on the hill where the Calpine lookout is situated. Between them, at Big Hill, the Bonta is absent and the Penman Formation rests directly on the old metamorphic rocks, yet just across the river on Mt. Jackson the Bonta is a thousand feet thick. These relationships indicate that the Bonta must once have been present over the site of Big Hill but that it was eroded away. Large faults separate the Sierra Nevada from the Plumas Trench and that from the Grizzly Ridge block. From all this one may infer that faulting after the deposition of the Bonta Formation resulted in the relative depression of the Grizzly Ridge block so that a great thickness of Bonta was preserved there; that a lesser depression of the Plumas Trench relative to the Sierra Nevada resulted in the preservation there of only a thin layer on parts of that block but permitted its complete removal from the area of Big Hill; and that a relative uplift of the Sierra Nevada block resulted in almost complete removal of the Bonta Formation from the area now occupied by the drainage basins of the Feather and Yuba rivers.

Here again we see evidence of an uplift of the Sierra Nevada by faulting, but that the uplift did not persist is obvious from the fact that erosion soon cut another relatively smooth surface across Grizzly Ridge, the Plumas Trench block, and the Sierra Nevada upon which the Penman Formation was deposited.

PENMAN FORMATION

Not only was the Bonta Formation exposed on the new surface eroded across it, but the old metamorphic rocks, the granitic rocks, and the rocks of the Lovejoy, Ingalls, and Delleker formations were again brought to light by erosion. Thus the Penman Formation deposited on the erosion surface rests on each of them at one or another place.

This erosion surface, like that upon which the Bonta Formation was deposited, seems to have been quite smooth. Although it is not easily studied it seems certain that there were no features of relief on it that could be called canyons or even deep valleys.

A topographic map made by T. H. Hemborg of a part of that erosion surface on the Sierra Nevada is shown here as fig. 108. This is the same area as that of figs. 85 and 86, and it is at the same scale and contour interval. It is also uncorrected for the tilt of the Sierra Nevada, and a few faults are not shown because their locations are not well known. The map shows a land of rolling hills 300 to 400 feet high bordering open valleys. The valley between Howland Flat and the neighborhood of Port Wine follows that of the Port Wine channel shown on fig. 85. This is a reflection of the ease of erosion of the Auriferous Gravel in contrast to the harder metamorphic rock on both sides of the channel. The valley near Gibsonville is about parallel to the La Porte channel but downstream it turns more directly to the west. The ridge between Gibsonville and Howland Flat is of hard metamorphic rock and has about the same form that it had when it was buried by the Auriferous Gravel some 50 million years before. It was exhumed, of course, from beneath the Auriferous Gravel. Interestingly, that same ridge, now called Sawmill Ridge, is there today, although Slate Creek cuts across it. It is also interesting that the relief and slopes, as reflected by contour spacing, are lower than they were at the time that the Auriferous Gravel was deposited. Obviously there was nothing like the modern canyon of Slate Creek.

An interesting detail of relief about a quarter mile west of the Howland Flat and south of the road, but partly obscured by trees, is a river channel cut into the Auriferous Gravel and filled with reddish pink mudflow breccia of the Penman Formation. It is exposed in the wall of the old hydraulic mine pit.

Like the Bonta Formation, the Penman consists mostly of andesite mudflow breccia and volcanic conglomerate, but there are sufficient differences between the two to permit a ready distinction in most places. The andesites of the Penman Formation are lighter colored than those of the Bonta; they are pink, pale gray, or light brown, often nearly white from a distance. The hornblende crystals are smaller and very much more abundant than in the Bonta rocks. In

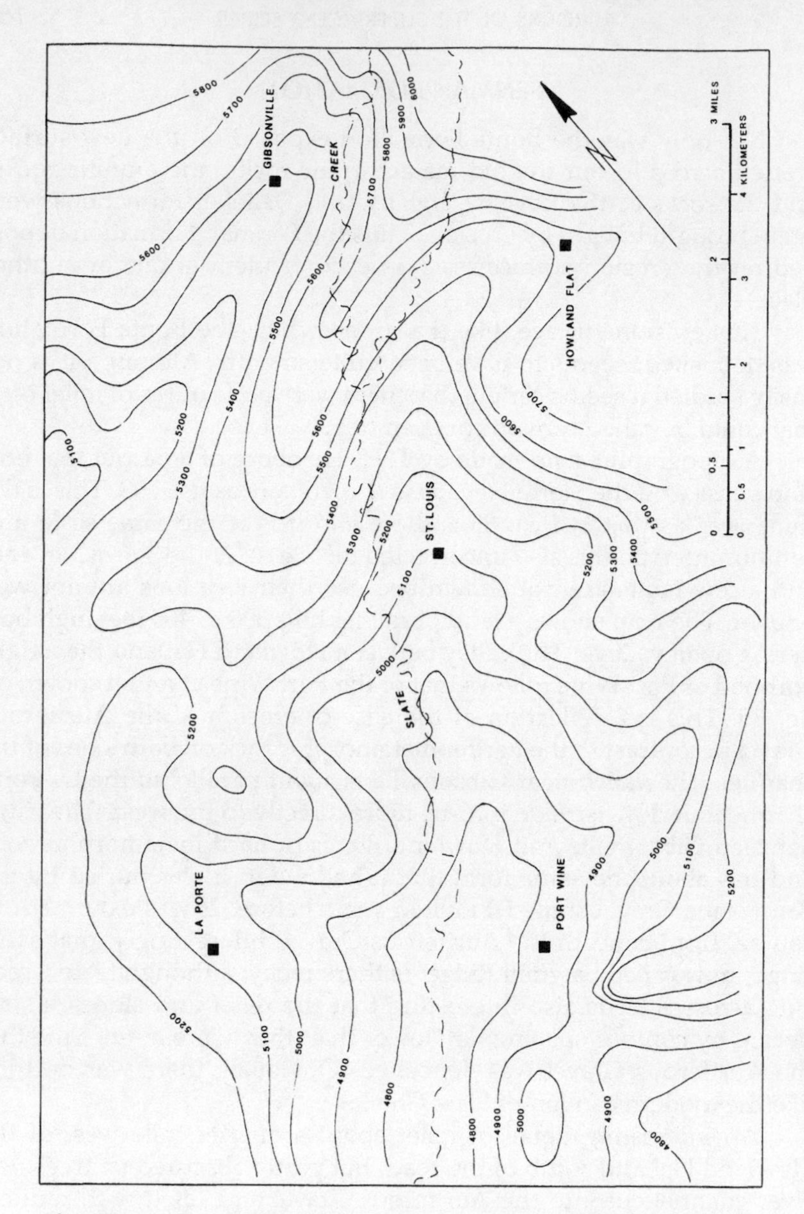

Fig. 108. A late Miocene landscape. A topographic map of the surface upon which the Penman Formation was deposited in the La Porte–Gibsonville District. Compare this to figs. 85 and 86, which show the same area at the same scale and contour interval during the middle Eocene and at present. From the unpublished studies of T. H. Hemborg.

many places the wheel tracks of unpaved roads glisten with sunlight reflected from bits of hornblende crystals. There is almost no foreign rock material—no volcanic ash eroded from the Delleker Formation, no dark andesite eroded from the Ingalls Formation, and no granitic debris. The soil over the Penman is gray to black and soft; it does not form reddish crusts as does that over the Bonta. Hillsides slope more gently so that where Penman rests on Bonta there is commonly a distinct change in slope at the contact. On south slopes especially, the forest is notably more thrifty than it is on the Bonta Formation.

Lava flows are known in two places: a few feet below the top of Mt. Fillmore and on the ridge southeast of Smith Peak.

The Penman Formation is widely distributed over the southeastern part of Grizzly Ridge, and around Mt. Ingalls. It is present in the Plumas Trench from near Quincy to Sierra Valley. On the Sierra Nevada it extends from the crest to the Sacramento Valley. To the south it probably merges into similar rocks named the Mehrten Formation. Its distribution to the north and east is not yet well known. There is a very little on Beckwourth Peak, and perhaps none in the mountains north of Sierra Valley, where it was probably deposited but eroded away at a later time. Similar rocks of about the same age are present in western Nevada.

Evidence for westward transport of the Penman rock is less abundant than for Bonta rock, but what there is is quite convincing. On the southeastern part of Grizzly Ridge, the lowest 25 to 50 feet of the Penman Formation contains blocks and chips of a gray to white flow-layered volcanic rock that has a very distinctive appearance (see fig. 26). It was evidently erupted from a vent, now a plug, of identical rock, about one mile southeast of Smith Peak. The pieces are found to the south, southwest, and west of there as far as Long Valley Creek. They are progressively finer farther from the obvious source. They are not present to the north or northwest, and they have not been found on Big Hill or the Sierra Nevada. The transport was clearly west and southwest as far as 8 miles from the source.

On the west side of McRae Ridge in the NW 1/4, Sec. 20, T. 22 N., R. 11 E., upslope from the Johnsville–La Porte Road, the base of the Penman Formation is marked by many slabs, some as large as 12 feet across, of meta-andesite that undoubtedly came from the Taylor Formation although the underlying rock is the Shoo Fly Formation (fig. 109). The least distance that these rocks could have traveled is 3.5 miles directly to the southwest. They could have come from the north or east, but the distance of travel would have been greater. Whatever the direction from within the northeast quadrant, they are either on or southwest of the summit of the Sierra Nevada, regardless of whether one prefers to think of McRae Ridge or Eureka Ridge as the

Fig. 109. Penman Formation. Large slabs of meta-andesite breccia of the Taylor Formation resting on the Shoo Fly Formation mark the base of the Penman Formation and indicate transportation in a westward direction. On the west side of McRae Ridge above the Johnsville–La Porte Road. NW 1/4, Sec. 20, T. 22 N., R. 11 E.

present crest. In any event, they show very clearly that there was no Sierra Nevada at that time.

Much of the andesite of the Penman Formation has many small vesicles, which permit it to weather easily. The mudflow breccia and volcanic conglomerate are soft and easily eroded. For these reasons, and the good growth of forest, the Penman rocks are not well exposed. However, a good place to see the Penman is in the road cuts on State Highway 70-89 where it passes through the narrow canyon of the Middle Fork of the Feather River between Blairsden and Cromberg. The same rocks are also well exposed across the river from the highway in the great scar just northwest of Camp Layman, shown in fig. 110. The Penman is also well exposed on the ridge southeast of Smith Peak (fig. 111), and, on the north side of Mt. Fillmore, an easy walk and climb from the Johnsville–La Porte Road where it passes through the saddle in the NE 1/4, Sec. 27, T. 22 N., R. 10 E.

An excellent place to compare the Bonta and Penman formations, although the actual contact is not exposed, is in Secs. 11 and 12, T. 24 N., R. 12 E., on the crest of Turner Ridge. This location, already mentioned as an excellent spot to observe the Bonta Formation, is reached

Fig. 110. Penman Formation. Light-colored hornblende andesite mudflow breccia on the west side of the Middle Fork of the Feather River at Camp Layman. The view is across the river from State Highway 70-89. SW 1/4, Sec. 19, T. 23 N., R. 12 E.

Fig. 111. Penman Formation. Mudflow breccia southeast of Smith Peak. The cliff is about 150 feet high. Sec. 16, T. 23 N., R. 13 E.

along the road that leads to Mt. Ingalls, about 2.5 miles from its junction with the Walker Mine Road on the north shore of Lake Davis. An abandoned road to Mt. Ingalls follows the crest and is crossed by the newer road at this point. All the open land along the ridge to the southeast, and also to the west, is on the Bonta Formation, and most of its distinctive features are visible here. To the northwest the no longer used road enters forest a short distance from the intersection. The lower limit of forest marks the contact of Penman on Bonta. About a mile farther along the road to Mt. Ingalls, in the NE 1/4, Sec. 11, the Penman Formation is exposed in road cuts, well enough so that its differences from the Bonta are obvious.

Like the Bonta Formation, the Penman Formation was eroded after deposition so that its original thickness cannot be known. On Grizzly Ridge northeast of Penman Peak the Penman is a little more than 1,000 feet thick, and on the Sierra Nevada at Mt. Fillmore it is 950 feet thick. There may be places not yet studied where it is even thicker.

The age of the Penman Formation is not known by radiometric dating because no determinations have been made. However, a satisfactory conclusion can be drawn even though some of the information is from remote places.

Fossil leaves found on Grizzly Ridge about a mile north of Blairsden are thought to indicate an age of about 10 million years. Fossil leaves in what is undoubtedly the westward extension of the Penman Formation at Remington Hill about 12 miles east of Nevada City are thought to be about 9 million years old. Near Mount Reba, about 12 miles west of Ebbetts Pass, fossil leaves are associated with hornblende andesite mudflow breccia that is dated radiometrically as 7 million years old. The rock is similar to that of the Penman Formation.

The Mehrten Formation of the foothills of the central Sierra Nevada also contains fossil leaves and the bones of larger animals. Two such occurrences, one near Knights Landing on the Stanislaus River and another on an island in Turlock Lake east of Modesto, are both younger than 9.5 million years, the radiometric age of the lavas of Tuolumne Table Mountain. By comparison with other occurrences in the San Francisco Bay Region, it has been suggested that these two may be as young as 4.5 to 5 million years.

All of these fossils and the rocks that contain them are younger than the Bonta Formation. Until recently they were considered to be Pliocene in age, but they are now relegated to the later part of the Miocene.[3] Their widespread occurrence lends support to the notion

3. Until 1975 the Pliocene epoch was thought to have lasted about 10 million years and to have ended about 2 million years ago. Since then, as a result of the study of the sedimentary rocks of the deep oceans, it has been restricted to only 3.5 million years, ending about 2.5 million years ago. Considerable confusion has resulted.

that the hornblende andesite of the west slope of the Sierra Nevada is of the age range of 10 to 5 million years, that the Penman Formation is roughly equivalent to the Mehrten Formation, and that if the older Bonta Formation was deposited on the Sierra Nevada proper it was removed by erosion before the younger rocks were deposited. However, the last conclusion concerning the Bonta is speculation based on negative evidence—that is, the absence of positive information. It may be proven wrong at any time.

The character of the country when the Penman Formation was deposited must have been about the same as when the Bonta Formation was deposited. According to D. I. Axelrod the nature of the fossil plants shows that the present summit region of the northern Sierra Nevada had an altitude of not more than 2,000 feet. The climate was somewhat cooler and drier than in Bonta time. This reflects a long-term trend toward cooler and drier climate that began in the Eocene and culminated in the glacial episodes described in Chapter 8.

The large-scale faulting that had preserved the great thickness of the Bonta Formation to the north and east and provided for its removal from the Sierra Nevada had produced no permanent elevation. The fossil plants of western Nevada show that no barrier existed during the time in which the Penman was deposited—there was no rain shadow—and there is the physical evidence that great blocks of the Taylor Formation rocks were transported westward onto or across the present summit region. Volcanoes were active, but no great cones resulted. Streams flowed westward from Nevada into the Sacramento Valley. The country was covered with vegetation and animal life was abundant.

INTRUSIONS OF ANDESITE

Intrusions of andesite are exceedingly numerous in the northern Sierra Nevada and adjacent regions. Some are dikes, but most are roughly oval or circular in plan, to which the name plug is appropriate; they are the once-magma-filled throats of long-extinct volcanoes.

Of the scores of such intrusions known, none seems to be of such composition that it could have been a source for the rocks of the Ingalls Formation. Many intrusions, especially those along the crest of the Sierra Nevada, but also some on Grizzly Ridge, occur in the Penman Formation. Because no lavas, mudflow breccias, or other deposits of andesite younger than the Penman are known, those plugs are presumed to have been sources for the Penman rocks; they cannot have been sources for the older Bonta Formation. However, most of the plugs intrude the Bonta Formation or older rocks and therefore cannot be assigned an age, except by inference, until radiometric dates are obtained. Those that are of light-colored rock with abundant small phenocrysts of hornblende are inferred to have been sources of Penman rocks. Darker-colored rocks with large hornblende crystals, and a few of the dark-colored rocks, andesite without hornblende, are presumed to have been sources of the Bonta Formation rocks.

Many intrusions are of massive rocks but some are breccia, and some are of both. Many are multiple intrusions, that is, they are composed of two or more separate intrusions. A few intrusions, mostly dikes, but also some plugs, have columnar structure. Some dikes, more resistant to erosion than the adjacent rock, stand out as walls, and those that are columnar resemble stacked cordwood (see fig. 6). A massive, nearly circular intrusion about 4,500 feet in diameter on the west side of Lake Davis, shown in figs. 112 and 113, is made of very large columns that are steeply inclined toward the margin of the body.

Some intrusions that are more resistant to erosion than surrounding rock stand out as prominent hills or mountains. Among these is the hill of massive andesite called Sugar Loaf (see fig. 27), which is north of State Highway 70 about 3 miles northeast of Beckwourth, and the craggy mountain of andesite breccia about 3.5 miles south of Beckwourth on the west side of the county highway to Calpine and Sattley. The narrow canyon of the Middle Fork of the Feather River east of Portola is cut through granitic rock and a massive but very strongly jointed intrusion of greenish andesite. Intrusive contact of the andesite against the rather badly decayed granitic rock can be examined in cuts along both the new divided highway and the old road at river level.

In the vicinity of Bunker Hill Ridge from Pilot Peak to the head of the West Branch of Nelson Creek, many plugs came through the Shoo Fly Formation and intruded the Penman Formation. What is perhaps the largest andesite intrusion of this entire region includes Blue Nose Ridge, the lower east slope of Mt. Fillmore, Mt. Etna, Mt. Stafford, and Beartrap Mountain. It is, of course, a complex intrusion, some of which is breccia and some of which is massive. Its southern limit is not determined but it occupies no less than 8 square miles.

The intrusions of breccia are of great interest because they may be direct sources for at least a part of the mudflow breccias of the Bonta and Penman formations. Many of them can be recognized by the fact that the forest on them is remarkably open or absent. Soil is totally absent from many, and steep-walled gullies in bare breccia are a common feature.

A good example is the small body shown in fig. 114 that is near the NW Cor., Sec. 9, T. 24 N., R. 13 E., 1,500 feet south of the intersection of the Bagley Pass Road with the Beckwourth–Genesee Road at the west end of Clover Valley. It is a nearly circular body, about 800

Fig. 112. Intrusion of columnar hornblende andesite, about 4,500 feet in diameter, on the west side of Lake Davis. Large columns visible on the left side slant down and outward. The vent now filled with light-colored rock probably was a source for material of the Penman Formation. Sec. 4, T. 23 N., R. 13 E.

Fig. 113. A closer view of the columns of the plug of andesite shown in fig. 112.

Fig. 114. A plug of brecciated light-colored andesite intruded into dark-colored mudflow breccia of the Ingalls Formation. Oval shaped and about 800 feet in diameter, it is crossed by the Bagley Pass Road near its junction with the Beckwourth–Genesee Road in the NW 1/4, Sec. 9, T. 24 N., R. 13 E.

feet in diameter, of light gray to cream-colored hornblende andesite that is in sharp contrast to the dark gray mudflow breccia of the Ingalls Formation that it intrudes. The close views of the rock shown in figs. 115 and 116 show well its nature: angular fragments of andesite set in a matrix of successively smaller fragments of the same rock down to powder size, without visible openings. All pieces of rock are alike. Traces of parallel sets of joints remain, but they disappear into the breccia—they are older than the brecciation. The rock is coherent to a degree; pieces of the breccia can be handled, but if lightly struck with a hammer they crumble. It seems obvious that this rock was brecciated in place. It cannot be rock thrown out of a volcanic vent that fell back into the vent and cannot possibly be a part of the wall of a crater that collapsed into the vent. It seems to have been broken by processes not yet understood.

Another easily accessible example is that shown in fig. 117, a view of the great, barren erosional scar on the west side of McRae Meadow in Secs. 29 and 32, T. 22 N., R. 11 E., about 4 miles east of Johnsville along the La Porte Road. The intrusion is nearly circular and about two-thirds of a mile in diameter. It intruded through the Shoo Fly Formation into the Penman Formation. The intrusion is multiple, consisting of not less than twenty distinct bodies. Figure 118, photographed from a central point, shows several of them. The contact of one intrusion breccia against another can be seen in many places; one is shown in fig. 119, which also shows a close view of the brecciated andesite. The margin of the intrusion is at the edge of the

Fig. 115. A close view of the brecciated rock of the intrusion shown in fig. 114. This view shows the remains of a set of joints developed in the cooling igneous rock before it was brecciated.

Fig. 116. A very close view of the brecciated rock of the intrusion shown in figs. 114 and 115.

Fig. 117. The andesite breccia intrusion on the west side of McRae Meadow in Secs. 29 and 32, T. 22 N., R. 11 E., 4 miles east of Johnsville via the La Porte Road. This complex brecciated body, nearly circular and about 3,600 feet in diameter, intrudes both the Shoo Fly and Penman formations. Without doubt, it was an eruptive center for the mudflow breccias of the Penman Formation.

forest except for the east side, which is concealed beneath McRae Meadow.

The rock of many intrusion breccias is green because of the presence of the green mineral chlorite, which is the product of the decomposition of hornblende or other iron-bearing minerals. The transformation is a common one that results from attack by hot water in an environment free of oxygen. No doubt the water was that which remained after the magma had crystallized.

Both the mudflow breccias and the blocks of andesite that they contain, in both the Bonta and Penman formations, show the effect of the availability of oxygen—that is to say, they are oxidized. This is indicated by the common pinkish, reddish, yellow, and brown colors, and by the fact that the hornblende in them is a dark reddish brown color. However, in a few places, notably in the mudflow breccia of the Bonta Formation, there are large blocks of green breccia, some more than 10 feet in diameter, like that which is so common in the intrusions (fig. 120). From this fact I conclude that some intrusion breccia came to rest in a vent, was altered, and then extruded by re-

Fig. 118. A view within the breccia intrusion at McRae Meadow shown in fig. 117. Although not clearly identifiable in the picture, at least ten separate intrusions are present in this field of view.

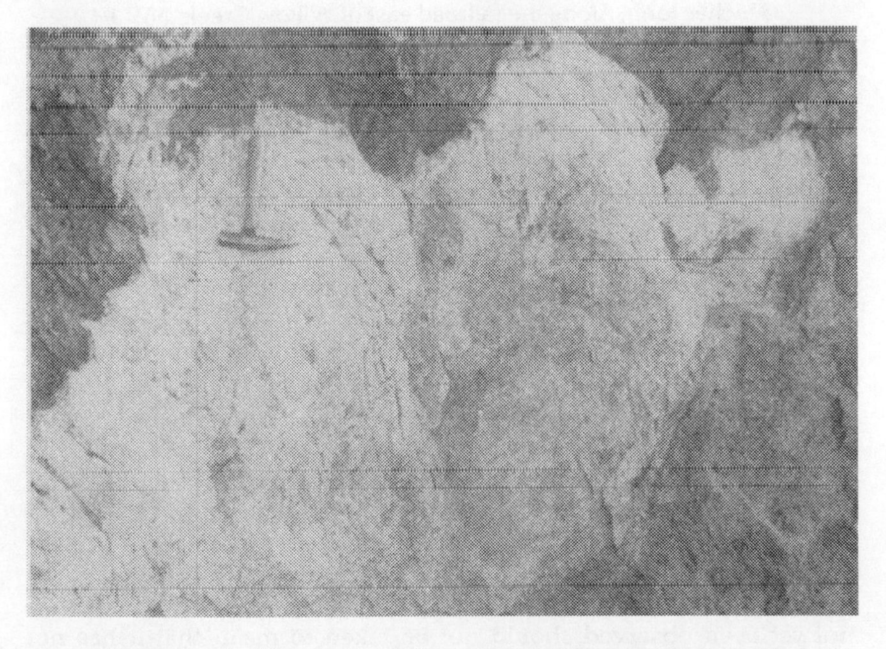

Fig. 119. A close view of the contact between two intrusive breccias of the McRae Meadow complex shown in figs. 117 and 118.

Fig. 120. An angular block of green andesite breccia in pink andesite mudflow breccia of the Bonta Formation. The two breccias look alike except for color. The hammer handle is 13 inches long. Along the railroad east of Willow Creek. NW 1/4, Sec. 19, T. 22 N., R. 13 E.

sumed activity. Some of the green breccia erupted in this way was in large enough pieces to escape oxidation. One would expect that smaller pieces would become oxidized along with the other blocks. The conclusion is inescapable that breccia was extruded and that the eruptions were not the usual ones in which pieces of rock are thrown violently from the throat of a volcano to fall into a loose, porous heap.

Some of the lava flows of the Bonta Formation in the region north of Sierra Valley and andesite lavas elsewhere in the Sierra Nevada show the same pattern of brecciation, which, therefore, is not confined to intrusions. This is not the pattern that results from the breakup of the crystallizing crust of a lava owing to continued movement of a still-fluid interior. We leave this problem in an unresolved state. The cause of the brecciation and the relationship of the breccia intrusions and lava flows to the mudflow breccias is uncertain. But the fact that the direct eruption of a muddy volcanic breccia has not yet been observed should not be taken to mean that it has not happened—neither has anyone yet witnessed the formation of a welded tuff.

WARNER FORMATION

What had happened many times before happened once again; after the Penman Formation was deposited, another episode of faulting followed by a period of erosion exposed at the surface, at one place or another, nearly all of the older formations. Erosion removed the relief that resulted from faulting so that another surface of unconformity, apparently a quite smooth one, came into being; it too extended across the site of the Sierra Nevada and the lands to the north and east. The lava flows of the Warner Formation were erupted on that surface from numerous vents.

Not much can be said about the details of the erosion surface on which the lavas rest. Much of it is concealed by brush, and by a screen of loose pieces of the lava that have moved down the slope, and not much of it is available for study because most of the Warner Formation has already been removed by erosion.

The Warner Formation is almost entirely lava flows. Only trifling amounts of sand and pebbles have been found between lavas; there is a little volcanic ash, and here and there masses of cinders.

The lava of the Warner is basalt but it is quite unlike that of the older Lovejoy Formation. The color is mostly medium to light gray. Well-developed joints parallel to the top and bottom of the flows permit the rock to split into flat pieces that range from slabs like flagstones to small chips. In sunlight the broad surface of a clean slab or a freshly broken surface parallel to the broad surface of a slab has a distinct sheen that is absent from the edges of such pieces. This is the result of the flow of the magma during crystallization; flow caused the alignment of the multitude of tiny tabular grains of feldspar that comprise the bulk of the rock. In this respect the rock somewhat resembles slate, but the igneous basalt does not have cleavage, and the cause of the orientation of the grains is quite different.

It is not easy to study individual lava flows or to count them except on cliffs because of that screen of loose rock already mentioned. However, the lava flows are thin, mostly in the range of 10 to 30 feet, and they are massive in the interior and scoriaceous at the top and bottom. These features, shown in fig. 121, can be seen along the road near the top of Mt. Ingalls. Columnar jointing is present, but is obvious in only a few places.

The most conspicuous mineral of the lavas is olivine, which, though normally a clear green color, is in these lavas pale yellow to golden yellow owing to slight alteration. The grains (phenocrysts) are a sixteenth to a quarter inch across. Less common than olivine is grass green to blackish green augite, either unaccompanied by olivine as it is in the rocks of Smith Peak, or with it as in those of the upper part of Mt. Ingalls.

Fig. 121. Warner Formation. A basalt lava flow that has a massive interior, but a slaggy top and bottom of scoriaceous fragments. On Mt. Ingalls by the road to the summit. NW 1/4, Sec. 34, T. 25 N., R. 12 E.

A very interesting aspect of much of the lavas of the Warner Formation and of the volcanic plugs that fed them is that they are contaminated with granitic debris. This came about because the granitic rock was exposed and weathered on each of the surfaces of unconformity older than the Warner. For example, at places where the granitic rock can be seen beneath the Bonta Formation it is like the material shown in fig. 5, an incoherent granular mass somewhat resembling sand.

This loose material evidently poured into the rising columns of magma, where they penetrated it on their way to the surface, for it is abundantly present in both lavas and plugs where the underlying basement rock is granitic rock. Figure 122, for example, shows part of a plug on the north side of Red Clover Creek near the SW Cor., Sec. 33, T. 25 N., R. 13 E., containing pieces of decayed granitic rock several inches in diameter. Elsewhere sheets of granitic material one grain thick were smeared along the flow layers of the basalt. Bits of granitic rock a fraction of an inch in diameter are common and easily recognized.

In some instances this contamination had a profound effect on

Fig. 122. A large piece of granitic rock and disaggregated grains of granitic rock in the basalt of a plug, one of many source vents for the lava flows of the Warner Formation. A small hill on the north side of Red Clover Creek near the SW Cor., Sec. 33, T. 25 N., R. 13 E.

the magma and the resulting igneous rock. The addition of the large amount of silica of the granitic debris caused the mineral olivine, a normal constituent of the basalt, to be either reduced in amount or absent. The mineral hypersthene, a member of the pyroxene group, occurs in its place. As a result of the contamination, some of the magma crystallized to andesite instead of basalt. The rock is still recognizable as belonging to the Warner Formation or the equivalent intrusions because it retains its other characteristic features.

Although the Warner Formation was widespread, its present areas of occurrence are small because erosion has already removed most of the lava. Much of that which remains is on high peaks and ridges although some occurs in the Plumas Trench and it is probably present beneath the Mohawk Valley.

In the Diamond Mountains on Thompson Peak some thirty lava flows have a total thickness of about 800 feet. Warner lavas are present in the hills east of Frenchman Reservoir, and at Portola they comprise the cliff-forming, columnar jointed rocks at the top of Beckwourth Peak. Lavas to a thickness of 1,000 feet make up the higher part of Mt. Ingalls. Lesser thicknesses are on the ridges adjacent to Smith

Fig. 123. Penman Peak from the top of the grade on County Road A-14 between Mohawk and Johnsville. The dark cap consists of basalt lava flows of the Warner Formation 350 feet thick. The rock of the light-colored erosional scar near the center of the picture and below the dark lava is mudflow breccia of the Penman Formation. The rock below that is all of the Bonta Formation.

Peak, the summit of which is a plug of the same lava. An undetermined number of flows totaling 350 feet in thickness make up the upper parts of Mount Jackson and Penman Peak (fig. 123).

In the Plumas Trench, State Highway 70-89 crosses a small patch of Warner marked by a very red soil a half mile east of Lee Summit, and another at its junction with the Sloat Road. State Highway 89 also crosses Warner lava at the southeast end of the Mohawk Valley, one and a half to two miles west of the pass to Sierra Valley, an area also marked by very red soil. The largest area of Warner in the Plumas Trench is that of Big Hill, where lava about 350 feet thick, the same as on nearby high-standing Mt. Jackson and Penman Peak, covers about 4 square miles.

State Highway 89 also crosses Warner lava on the grade south of Sierraville, and almost continuously from the Little Truckee River to the town of Truckee.

The Warner Formation is also present on the Sierra Nevada. The lavas are present at the summit of Claremont Hill south of Quincy and in numerous fault blocks east of the summit as far as the pass on the Quincy–La Porte Road between Thompson Creek and Nelson Point. A thin lava flow is present on the crest southwest of Gold Lake, and several areas of the basalt are present on the ridge between the North and Middle forks of the Yuba River, south of Downieville and Sierra City.

These scattered occurrences are the remnants of a once continuous sheet of lavas that covered most if not all of the Diamond Mountains, the Plumas Trench, and the eastern part of the northern Sierra Nevada.

The known extension of these rocks southward to Truckee and Lake Tahoe and the presence of similar rocks far to the north and east may mean that what is described here is only a small part of a much more extensive sheet of volcanic rocks, but almost nothing is yet known about the age relationships between the rocks at various places.

Similar rocks are not known in the western part of the northern Sierra Nevada, nor are they present under the Sacramento Valley. It may be that they never extended so far.

Source vents for the Warner lavas are abundant in the crestal region of the Sierra Nevada and in the Diamond Mountains. Most of them appear to be about circular but those whose borders can be closely examined are not so regular in form. Many are oval, or even angular, and some have tongues extending outward along faults. Many are obviously located at the intersections of faults. Their diameters range from a few tens of feet to 3,000 feet.

Plugs that can be easily reached include Smith Peak north of Portola; the summit of Mt. Ingalls, which at the lookout tower is an exceptionally coarse-grained rock; the hill a half mile southwest of Claireville townsite (see fig. 148) that stood as an island in Mohawk Lake, and the craggy black rock on the west slope of Mt. Jackson (see fig. 107) visible from the Jackson Creek Road. Another is on the east side of McRae Meadow at the junction of the roads in Sec. 21, T. 22 N., R. 11 E.

A most interesting and accessible plug is that at the bridge across the Middle Fork of the Feather River at Camp Layman between Cromberg and Blairsden, shown in fig. 124. Here the river has cut through a plug that is about 2,000 feet in diameter. The outer part of the plug as seen along the south side of the river is a pale gray rock that breaks into thin slabs. The lighter-colored central part of the plug above the river is a vent breccia—fragmented rock that filled the vent at or below a crater. Beneath the north end of the bridge is basalt with well-developed columnar joints (fig. 125) that may be a massive central part of the plug or a separate intrusion. Most of the plug is concealed be-

Fig. 124. A view of the interior of a plug of basalt, one of the many feeders for the lava flows of the Warner Formation. The plug is oval shaped and about 2,000 feet in diameter. The light-colored triangular area at the center of the picture is evidently a vent breccia—that is, fragmental material that fell back into the throat of the volcano. The bulk of the plug is microscopically flow layered, and for that reason it breaks into flat slabs and chips. At the south end of the bridge across the Middle Fork of the Feather River at Camp Layman. Sec. 30, T. 23 N., R. 12 E.

neath the river, alluvium, and a large landslide on the northeast side of the river.

The age of the Warner Formation remains a problem. No fossils have been found in it, and no radiometric age determinations have been made on rocks in Plumas, Sierra, or Lassen counties. Obviously the lavas are younger than the Penman Formation on which they rest, and they are older than the Mohawk Lake Beds described farther on. They are also older than the uplift of the Sierra Nevada for they are present at high elevation on the Sierra Nevada and several thousand feet lower in the adjacent Plumas Trench. They are older than the deep canyons of the North and Middle forks of the Yuba River because they are perched on the divide between them.

Several radiometric ages of similar rocks in the Truckee-Tahoe region, determined by the potassium-argon, or K-Ar, method are be-

Fig. 125. Basalt with columnar jointing. Probably a minor intrusion into the plug shown in fig. 124. Beneath the north end of the bridge at Camp Layman.

tween 2.5 and 1.2 million years, which is latest Pliocene and Pleistocene. Another K-Ar date on similar rocks at Verdi, Nevada, is 11 million years, and another north of Virginia City, Nevada, is 6.9 million years. Some of these dates may not be valid, however, because some of the dated lava flows had been contaminated by granitic debris—the significance of which is explained in Chapter 5.

Similar lavas were erupted in very recent time in the Owens Valley and in northeastern California, so it is conceivable that radiometric dates on this kind of lava that range from 10 or 11 million years to very recent are all valid. But, of course, not all of these rocks, regardless of their similarity, are necessarily to be included in the Warner Formation as it is defined in eastern Plumas County.

In the absence of more accurate information it appears that the Warner Formation in the northern Sierra Nevada and Diamond Mountains is late Pliocene in age, and it is possible that eruptions continued into the Pleistocene. Thus it might extend through a time span of perhaps 3 million years, and the youngest lavas could be 1 to 2 million years old.

During and at the end of the course of eruption of the Warner Formation the region was perhaps studded with cinder cones that marked the vents from which the lavas poured. None remain entire, but a part of one is embedded in the lavas on the north side of Mt. Ingalls, as shown in fig. 21, and at some vents red cinders were preserved by being drawn down into the vents as the magma receded. The landscape probably resembled that which can be seen northeast of Mt. Lassen, or that present near Medicine Lake and the Lava Beds National Monument in Siskiyou and Modoc counties, or that near Little Lake in the southern end of the Owens Valley. No volcanoes are now recognizable, although it is possible that the highest part of Mt. Ingalls was part of one. The lava there is thicker than elsewhere nearby, and several vents, or plugs, are present near the summit, two of which are filled with fragments of scoriaceous lava. However, the form of Mt. Ingalls is not now that of a volcano because it has been subject to later faulting and extensive erosion.

Although the land surface was paved with lava, it was not everywhere a barren wasteland of rock and cinders. Local direct evidence of the nature of the climate is lacking, but it was perhaps a bit cooler and drier than when the Penman Formation was deposited; this was a continuation of a trend that began back in the Eocene. The altitude was not much, if any, higher than it was in Penman time. Even though it might take some hundreds of years for coniferous forest trees to become well established on lava flows, the few millions of years available would provide ample time for that to happen over and over again. No doubt there was grassland, woodland, dense forest, and the water-loving plants that grew along streams. Deciduous broad-leafed trees were dominant, but the giant Sierra redwood and other conifers similar to those now in the region were present on the higher ground.

Judging from fossils found elsewhere in the west, the animal population was large and diversified, but more modern in aspect than that when the Bonta and Penman formations were deposited. Camels, elephants, large cats, horses, and others whose descendents are now present only in the Eastern Hemisphere were important elements of the scene.

After the volcanic activity that produced the Warner Formation had ended, erosion once more began its ceaseless work and cut a surface of low relief across all the land of the Feather River country and far beyond. The Warner Formation and all older rocks were exposed on that surface, a surface that is important to subsequent history and is named, described, and discussed in Chapter 7.

SUMMARY

We are now near the end of the story of the Superjacent Series; there remains only the Quaternary period, which, however short, was a time of major geologic events. We now summarize and draw some further conclusions about the Superjacent Series deposited during the Tertiary period.

The main theme is the laying down on a basement that consisted of metamorphic and granitic rock a series of formations originally, and even now, in a nearly horizontal position. The marine Chico and "Dry Creek" formations and the Ione Formation belong essentially to the Sacramento Valley. Of the seven Superjacent Series formations of the Sierra Nevada and Diamond Mountains, all but the oldest, the Auriferous Gravel, are volcanic. Each of them is separated from the earlier one by an unconformity. Faulting occurred between each of the last six, and erosion leveled whatever relief had resulted from either the faulting or the depositional processes.

As a result there is no place in the Sierra Nevada or the Diamond Mountains where one can see the seven formations piled up in sequence from oldest to youngest. Only partial sequences remain.

The several episodes of faulting and erosion makes it possible for each younger formation to rest at some place on each formation older than it is, as well as on basement, and it is a fact that all such possible relationships—twenty-eight of them—exist in Plumas County. Thus, the disposition of the seven formations is in many respects a great jumble of blocks in which sequences are in many places different from those on the opposite side of faults. Examples of this are shown on the cross section of fig. 129, section B–C.

Of course the jumble is not entirely without order. Since the rock layers are still nearly horizontal, older formations do not rest on younger formations. The river channels of the Auriferous Gravel on the Sierra Nevada, although eroded and faulted, are still relatively intact. They are cut off at the crest of the Sierra Nevada, but bits occur on Grizzly Ridge and elsewhere in the Diamond Mountains. The broad valley down which the lavas of the Lovejoy Formation flowed can be approximately outlined.

The Ingalls Formation seems to be restricted to the region bounded by a line from Sloat to Clover Valley and east to the road from Portola to Lake Davis. If it was deposited on the Sierra Nevada, no remnant has yet been recognized. The Delleker Formation, much thinned by erosion, occurs in rather widely distributed areas, but that fact alone is enough to lend credence to the idea that it together with the Valley Spring Formation was once a continuous sheet over not only the northern and central Sierra Nevada, but also over lands far to the east.

The Bonta Formation, widespread to the north and east, is almost absent from the Sierra Nevada, yet there is reason to believe that it was deposited there, and was later eroded away. The preservation of a thick Bonta Formation east of the Sierra Nevada indicates a probable displacement along the faults at the eastern edge of the Sierra Nevada—a relative movement, with the Sierran side up and the other side down. The same situation existed with respect to the Penman and Warner formations. Even though the Penman is thick at Mt. Fillmore, it is absent a few miles away, and the Warner lava rests on basement. The Warner lava in turn seems to have been deposited over much of the northern Sierra Nevada and then eroded away.

Now it seems that although the eastern limit of the Sierra Nevada may have been established by faulting before the Ingalls Formation was deposited, no significant relative uplift of the northern Sierra Nevada occurred until after the Bonta Formation was deposited in late Miocene time. Any relief that resulted from faulting then or after the Penman and Warner formations were deposited was removed by subsequent erosion. No barrier to the westward flow of streams or to the landward passage of storms resulted. Nothing like the present Sierra Nevada had yet existed.

The present landscape, including the Sierra Nevada uplifted to its existing height, is the result of events that occurred during the Quaternary period. Those events are the subject of the next two chapters.

7

Origin of the Landscape

I. FAULTING, EROSION, AND EXTINCT LAKES

THE QUATERNARY PERIOD

It is generally agreed that the Quaternary period lasted about 2 million years, only half a percent of the 400 million years spanned by this story, yet two very important events took place during that short interval. One was the uplift of the Sierra Nevada, and the breakup of the lands to the east, an event that affected a large portion of the continent. It resulted in the fault block mountain ranges and valleys of the Great Basin and Diamond Mountains, and the cutting off of drainage systems east of the Sierra Nevada from the Pacific Ocean. The second great event was the appearance of glaciers, the culmination of a worldwide long-term progression toward cooler climate. Glaciation had a profound effect on the Sierra Nevada.

Both events resulted in very important changes in the landscape, and accordingly, the thread of this story now shifts from the recital of geologic history recorded in formations, to a consideration of history recorded, in large part, in the landscape. There were, however, important bodies of rock deposited during this time.

The absolute time of beginning of the Quaternary is currently based on events recorded in the rocks and sediments of the deep oceans. Because it is not yet possible to relate those events with any precision to events in the northern Sierra Nevada, I have chosen to begin the story of the Quaternary period at an unspecified time after the eruptive period of the Warner Formation was over and before the Old Erosion Surface, defined further on, was cut across those same lavas.

Events in the northern Sierra Nevada and Diamond Mountains

during the Quaternary period are not well dated, and their sequence is difficult to establish, partly because they occurred in different places and partly because they are not related to rocks deposited in a sequence; there are gaps in the record. This is a somewhat unsatisfactory state of affairs, but there is no help for it at present.

FAULTS AND FAULTING

Six episodes of faulting occurred during the Tertiary period. The most continuous, or what may be thought of as the main faults, trend northwesterly. Less continuous but not less important faults trend northeasterly. The spacing (or abundance) of faults is indicated in figs. 129 and 136. Their pattern is shown in fig. 136. There is reason to believe that the spacing and pattern shown in these figures prevails throughout the Diamond Mountains, in the Plumas Trench, and in the summit region of the Sierra Nevada. The offsets on most of the faults measured vertically is in the range of a few tens to a few hundreds of feet.

When faulting began again in the Quaternary period, the movements were on the already-existing faults, but on fewer of them, and the offsets became very much larger, measurable in thousands of feet on some faults.

This is how the basin-and-range structure developed; the country became divided into larger blocks, bounded by faults with larger displacements. In that manner the landscape was profoundly reorganized. The long-existing system of streams by which the interior drained to the Pacific Ocean was cut off. Some blocks became elevated to become distinctly mountainous areas such as Grizzly Ridge and Dixie Mountain, whereas others, such as Sierra Valley and Mohawk Valley, were relatively depressed to become basins that held lakes.

Where the details are known, the boundaries of the blocks are complex. For example, five faults are present on the northeast face of the Sierra Nevada between Mohawk Saddle and the floor of Mohawk Valley, and four faults are present on the southwest face of Grizzly Ridge between Mt. Jackson and the Middle Fork of the Feather River at Two Rivers (see fig. 131). These mountain fronts are not single fault steps, they are more like giant staircases. The fault steps constitute transitions between high-standing and low-standing blocks.

Similarly, in map view, the ends of blocks elongate northwesterly are not terminated by single, large, northeast-trending faults but by alternations between the two sets of faults, as is clearly shown on fig. 136 along the north side of Grizzly Valley. The blocks are not bounded by single faults, but by complex systems of faults. For this reason fig. 126 is not a fault map; it shows only the approximate limits

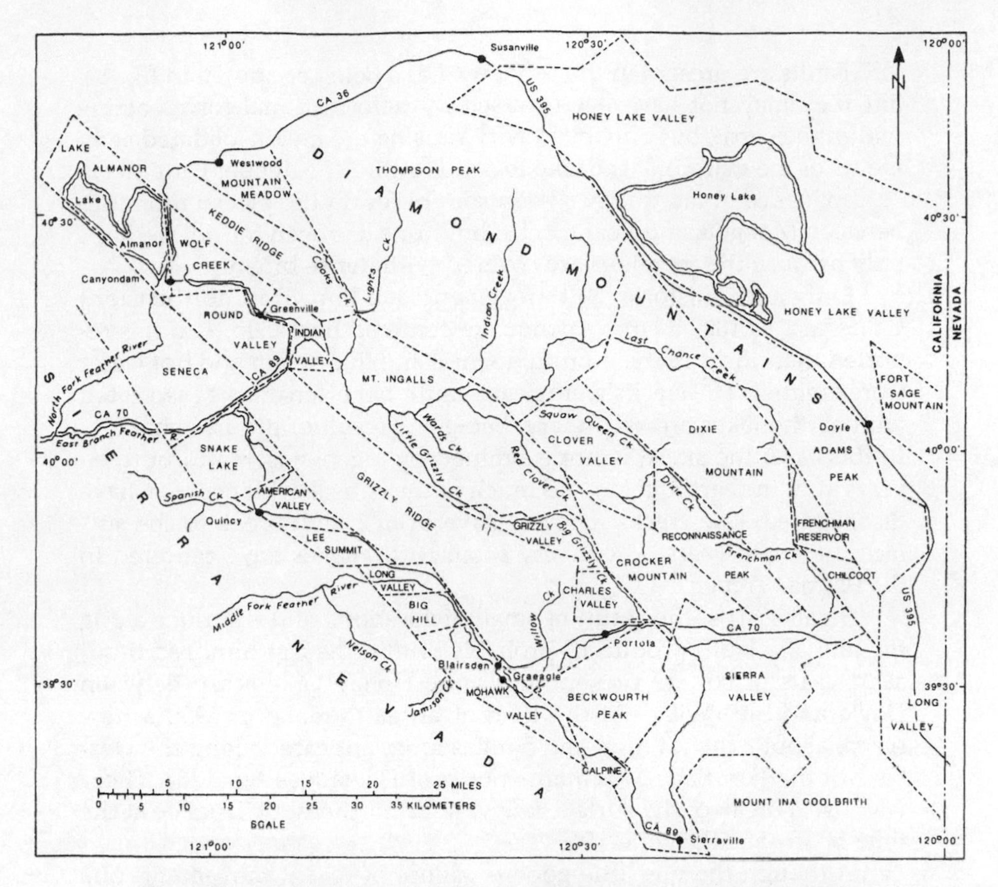

Fig. 126. A map showing the Sierra Nevada and the principal fault blocks of the terrain to the northeast. This is not a fault map. Although some lines are faults and some represent the several faults of a system, many are only the approximate boundaries between high- and low-standing areas whose differences of altitude are, without doubt, the result of faulting. Many boundaries may be changed as knowledge accumulates.

of relatively high- and low-standing blocks. Furthermore, the divisions shown on fig. 126 are provisional. The larger blocks will certainly be divided into smaller units as knowledge of the region accumulates.

The age of relatively recent faults cannot be stated with any degree of precision, but some crude estimates can be made. Faults cut the Warner Formation, which is probably no more than 2.5 million years old. They also cut the Old Erosion Surface, described further on, which is cut into the Warner Formation and is, therefore, somewhat younger than the faults. Faults that cut both the Warner and the Old Erosion Surface on Bill Hill pass beneath the Mohawk Lake Beds and seemingly do not cut the lake sediments.

Faults are present in the Mohawk Lake Beds, as shown in fig. 12, but they may not have been caused by deformational forces acting within the earth, but only by gravity causing the unconsolidated sediments of the margin of the lake to slide toward the deeper center.

Evidence of faulting has been sought for on the glacial moraines between Mohawk and Bear creeks, but none has been found, perhaps only because the moraines are covered with dense brush.

Only a few historic fault movements are known in northeastern California. In 1875 a large earthquake centered near Clio, and it is reported that cracks in the ground opened and that steam and hot water were emitted. The cracks were near a fault, but often cracks associated with earthquakes are secondary, being the result of ground shaking. In this case, the site was not examined by a geologist until fourteen years after the earthquake, and much of the critical evidence may have disappeared. The cracks may not have been a fault break to the surface. In recent years, a few very small earthquakes have centered in the Plumas Trench.

Small scarps, the result of small dislocations of the surface along the line of a fault, produced, probably within the last hundred thousand years or so, are present along the Honey Lake scarp between Doyle and Susanville. At the time of an earthquake in 1950 a new scarp, about 8 inches high and 5 miles long, appeared along the west base of the Fort Sage Mountains north of Doyle (see fig. 126). There was also a break of the surface a few miles northeast of Truckee at the time of an earthquake in 1966.

Although there is little good evidence of recent movement, one should accept the idea that the uplift of the northern Sierra Nevada is probably continuing, as is the case farther south in the Owens Valley, where a 20-foot fault movement occurred in 1872.

Faults can be obvious where exposed in road cuts, as shown in fig. 12; or along scarps produced at the time of earthquakes, such as that at Fort Sage Mountain in 1950; and in regions of rock that is bare because of aridity or glaciation. But in a forested and brushy country, faults are rarely visible. Like the great folds discussed in Chapter 4, they must be "worked out" by making a geologic map. To greatly oversimplify the matter, geologists walk through the country following the boundaries of the different formations, recognized as different kinds of rock, and draw them in their proper position on a topographic map or aerial photograph. In this region, where the formations are nearly horizontal, lines that define nearly horizontal surfaces represent the surfaces of contact between formations. Lines that define steeply inclined surfaces are most likely faults. If, for example, an area of Bonta Formation lies adjacent to or abuts against an area of Delleker Formation, the surface between them is a fault. Difficulties arise in areas covered with brush, or forest with thick duff on the floor, or areas of deep soil or of landslides and igneous intrusions, but there are ways to cope with such matters.

What geologists find out about faults in this kind of study is their direction on the ground, the direction and amount of their inclination, and the amount of offset of formations measured in the vertical direction. Rarely, if ever, is the "net slip" discovered—that is, the real direction and amount of movement measured on the fault surface. Furthermore, the motion discovered is relative motion—that is, how one side moved with respect to the other. Rarely, if ever, is a measure of the absolute motion determined—that is, how any block moved with respect to sea level. That requires information of a different kind and is discussed further on.

Thus the faults appear on geologic maps, and a representative sample of good maps has been made of the basin-and-range part of the Feather River country. The maps show that the blocks on fig. 126 are bounded by faults, that the differences in altitude between blocks is the result of movement on the faults, that the principal faults are normal faults (see fig. 36*c*), and that the crust of the earth was stretched, or expanded, or extended to accommodate the motion.

More specifically, geologic maps provided the information necessary to construct geologic cross sections, such as fig. 129 A–B–C. That section shows the gentle southwestward slope of the Sierra Nevada, and the essentially parallel position of the formations of Tertiary age. It also shows that the formations projected northeastward at the summit (Eureka Ridge) would pass out into the air, but that their continuation is at a lower altitude on Big Hill. Thus, geologic maps are fundamental because they contain the information necessary for understanding geologic structure—in this case, basin-and-range structure or block faulting.

A feeling for the fault block character of the region can be obtained by looking up and down the Mohawk Valley from its center. To the northwest, as shown in fig. 127, the valley ends against the higher-standing Big Hill block, whereas to the southeast, as shown in fig. 128, it ends against the Calpine block. On the southwest side is the Sierra Nevada, and on the northeast is the Grizzly Ridge block. The relative "lowness" of the Mohawk Valley block is spectacularly evident from such viewpoints as Mills Peak, Eureka Ridge, Eureka Lake or Eureka Peak, and the Bonta Ridge Road. Even the view into the valley and across it from State Highway 70 at the top of the grade about 2 miles east of Blairsden is impressive.

Fig. 127. A view to the northwest in Mohawk Valley. The ridge in the center is Big Hill, the Big Hill block of fig. 126. The skyline on the left is that of Eureka Ridge, the crest of the Sierra Nevada. The higher land to the right is Grizzly Ridge. The view is from the high point of land in the SE 1/4, Sec. 23, T. 22 N., R. 12 E.

THE OLD EROSION SURFACE

Following the volcanic episode that produced the Warner Formation, erosion cut a new surface across the Warner and all older rocks as well. Because a name is needed for purposes of description and discussion I am calling it informally the Old Erosion Surface. It is not geologically old—it is young—but I call it old because it is older than the uplift of the Sierra Nevada, the erosion of the deep canyons of the major streams, and the erosional carvings of the glaciers in the summit region.

The surface is one of rolling hills, of relief mostly less than 500 feet, although a few peaks stand above it. It is illustrated in profile in fig. 129, sections A–B and D–E, and in plan by the contour map of fig. 130.

On fig. 130 the surface is those areas between the South, Middle, and North forks of the Feather River, and elsewhere, where the contours are widely spaced. The principal stream courses incised into the surface are emphasized by heavy lines. Their canyon walls are the areas in which contours are closely spaced. The closer the contours

Fig. 128. A view to the southeast in the Mohawk Valley. The convexly upward slope on the right is the Mohawk fault scarp, the top of which is the crest of the Sierra Nevada. The low, nearly level ridge in the center is the Calpine block of fig. 126, and the hill near the left edge is the site of the Calpine lookout tower. The ridge in the far distance is east of Sierra Valley. The view is from the moraine ridge on the east side of Frazier Creek in the NW 1/4, Sec. 34, T. 22 N., R. 12 E. For views across Mohawk Valley see figs. 132, 149, and 150.

are, the steeper is the slope they represent. The profile of fig. 129 D–E shows the canyon of the Middle Fork incised to a depth of 3,000 feet below the Old Erosion Surface. Obviously the surface is older than the canyons cut into it.

The areas between the canyons seem to stand like plateaus, but they are not level; rather, they slope southwesterly to merge into the Great Valley. Eastward they seem to end at the crest of the range.

The slope of the surface in the Sierra Nevada is essentially the same as that of the formations of Tertiary age, and like them it did not end at the crest but is faulted down. Beyond the crest, the surface is recognizable as that on the Big Hill and Calpine blocks in the Plumas Trench (see fig. 126). Farther east it is found on Grizzly Ridge, around Clover Valley, and elsewhere in the Diamond Mountains as far as the Honey Lake fault scarp.

Southwest of Quincy the erosion surface on the Sierra Nevada ex-

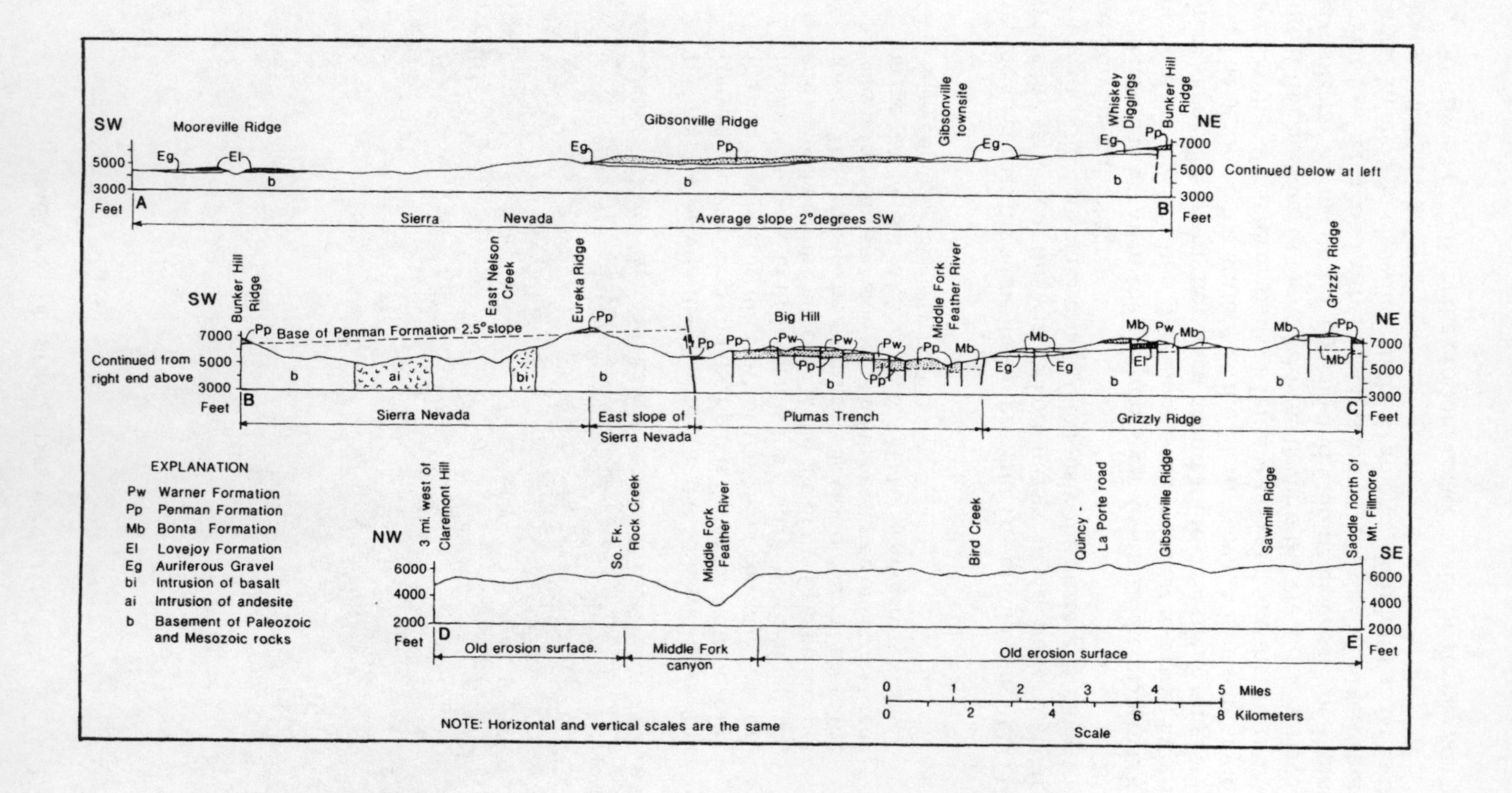
SW
Mooreville Ridge
Gibsonville Ridge
Gibsonville townsite
Whiskey Diggings
Bunker Hill Ridge
NE
Continued below at left
Sierra Nevada
Average slope 2°degrees SW
Feet
SW
Bunker Hill Ridge
East Nelson Creek
Eureka Ridge
Base of Penman Formation 2.5°slope
Big Hill
Middle Fork Feather River
Grizzly Ridge
Continued from right end above
Sierra Nevada
East slope of Sierra Nevada
Plumas Trench
Grizzly Ridge
EXPLANATION
Pw Warner Formation
Pp Penman Formation
Mb Bonta Formation
El Lovejoy Formation
Eg Auriferous Gravel
bi Intrusion of basalt
ai Intrusion of andesite
b Basement of Paleozoic and Mesozoic rocks
NW
3 mi. west of Claremont Hill
So. Fk. Rock Creek
Middle Fork Feather River
Bird Creek
Quincy - La Porte road
Gibsonville Ridge
Sawmill Ridge
Saddle north of Mt. Fillmore
SE
Old erosion surface.
Middle Fork canyon
Old erosion surface
NOTE: Horizontal and vertical scales are the same
Miles
Kilometers
Scale

tends to the summit of Claremont Hill, where it is on lavas of the Warner Formation. Between Claremont and American Valley, at least five steps bring the surface thousands of feet lower, almost to the level of the valley itself. Although the geologic map of that region is not complete, there is sufficient evidence to show that each step in the erosion surface is limited by a fault, simply because lavas of the Warner Formation are present on each step.

The knowledge that faults cut the erosion surface can be used to locate faults and obtain at least rough estimates of the vertical offset on some of them. Thus, the erosion surface can be used in the same way that the rocks of a formation are used, although in a far less precise way. It provides valuable knowledge in areas where the formations of Tertiary age have been entirely removed by erosion, as, for example, in the region northwest and north of Quincy.

There are complicating factors concerning the erosion surface. It is not in the same condition as it was before the block faulting began because erosion continues to act on it. Much of it has been removed by canyon cutting by modern streams. The surface is complex because different kinds of rocks weather and erode at different rates, other things being equal. But other things are not equal. Weathering and erosion rates vary with climate, and climate varies with altitude and time. So the Old Erosion Surface has been subject to varied attacks, culminating in that by glaciation in the summit region.

The Old Erosion Surface is very extensive, and its limits are not known. It does exist on the Sierra Nevada as far south as the Kern River. In the Yosemite region and beyond, where detailed studies have long since been made, several erosion surfaces have been recognized

---

Fig. 129. Sections A–B and B–C are together a geologic cross section showing the rocks of the Superjacent Series across part of the Sierra Nevada and Big Hill, in the Plumas Trench, to the summit of Grizzly Ridge. The location of the section is shown on figs. 79 and 80. Note the regional westward slope of the Sierra Nevada, which is the result of uplift and tilt, the low position of the Plumas Trench, including Big Hill, relative to the Sierra Nevada and Grizzly Ridge blocks, and the faulting internal to the Big Hill and Grizzly Ridge blocks.

The profile of Section A–B and of Big Hill is that of the Old Erosion Surface. The profile D–E, the location of which is also shown on figs. 79 and 80, is that of the Old Erosion Surface taken in a northwesterly direction to show the 3,000-foot-deep canyon of the Middle Fork of the Feather River incised into it.

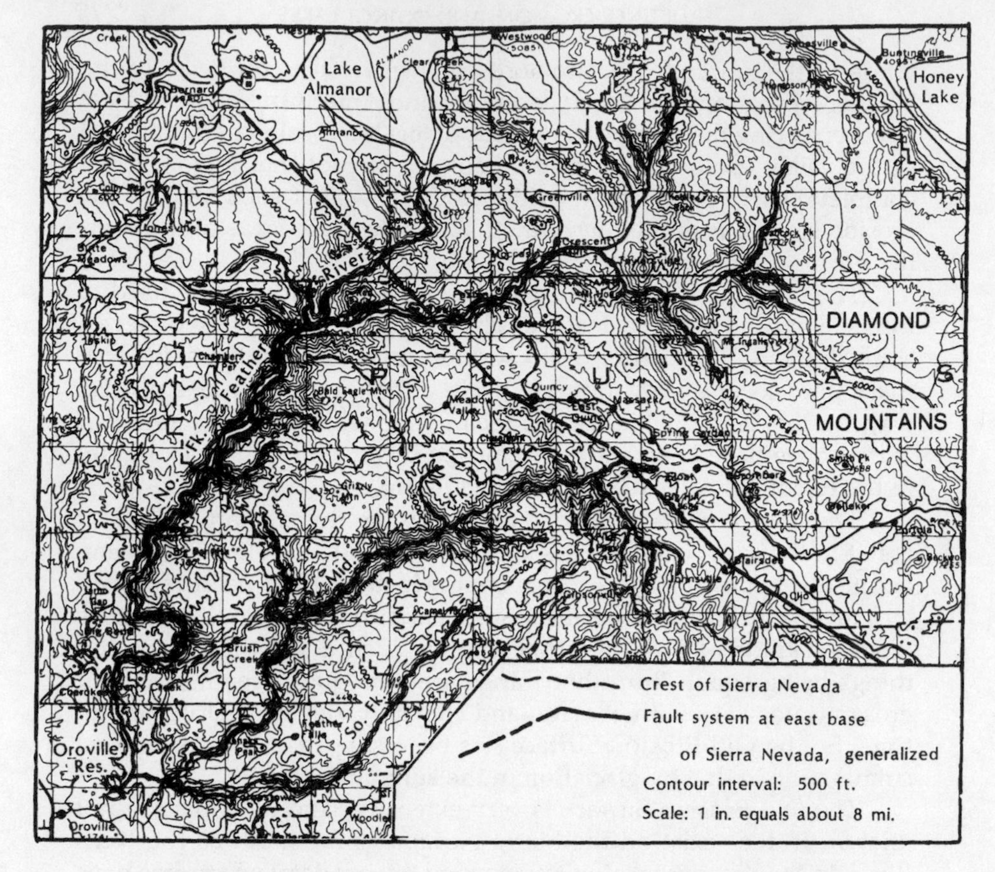

Fig. 130. The Old Erosion Surface. A contour map of the northern Sierra Nevada and Diamond Mountains. The Old Erosion Surface is the area of widely spaced contours except for the basins of the extinct lakes shown on fig. 38. It occupies the space between the canyons of the main streams of the Sierra Nevada, and most of the Diamond Mountains.

The courses of the main streams and their tributaries that are incised into the Old Erosion Surface are indicated by heavy lines. The steep walls of the incised canyons are marked by closely spaced contours congruent to the stream courses. Canyon cutting east of the Sierra Nevada is clearly shown.

that reflect stages in the uplift and glaciation of the range. Perhaps what I am treating as one surface may be also found to be more complex when studied in detail.

The Old Erosion Surface can be visited at many places. The road from Marysville to La Porte and on to Gibsonville townsite and Onion Valley is almost everywhere on it. Bucks Lake and Little Grass Valley reservoirs are on it and can be reached by good roads. Big Hill can be visited by way of the Poplar Valley Road and can be viewed from above from Eureka Ridge or Mills Peak. Clover Valley and other parts of the Diamond Mountains are accessible by relatively good roads.

UPLIFT OF THE SIERRA NEVADA

The uplift of the Sierra Nevada has been a matter of conjecture and debate for a long time. Early attempts to estimate the uplift were based on the Auriferous Gravel channels in the present basins of the Yuba and American rivers. It was accepted then that the summit region was about where it is today, that the relief of the range was about as rough as it is today, and that the channels were not disturbed by faulting and tilting. None of these assumptions is true, so the results of attempts to restore the ancient river beds to their original condition are inconclusive. Any valid restoration of those ancient river beds will require far more detailed information than is available at present.

Evidence from the Bonta Formation in Plumas County is instructive but does not yield a complete answer. The fossils near Gold Lake on the Sierra Nevada and those near Clio on the block east of the Plumas Trench, both close to the base of the Bonta Formation, are of plants that evidently lived at an altitude of no more than 2,000 feet, and possibly even lower. The altitude of the Gold Lake occurrence is now 6,500 feet, which indicates an absolute uplift of 4,500 feet. That at Clio is at an altitude of 4,400 feet, which indicates an absolute uplift of about 2,400 feet. Thus, the Sierra Nevada was uplifted with respect to a nearby part of the Basin and Range Province to the east by about 2,100 feet. Both areas were uplifted, but the Sierra Nevada was uplifted more than the region east of it.

Consideration of the Penman Formation yields another value. The base of the Penman in the summit region is at various altitudes, but nowhere is it far from 7,000 feet. Along lower Jamison Creek and the Middle Fork of the Feather River, both the Warner basalt and the Penman Formation can be seen to pass beneath the younger sediments of Mohawk Lake. If we assume that beneath Mohawk Valley near Graeagle the Mohawk Lake Beds are 750 feet thick, a reasonable figure, that the Warner is 350 feet thick, as it is on Mt. Jackson and

Penman Peak, and that the Penman is beneath it and is 1,000 feet thick, the same as it is on Grizzly Ridge, the base of the Penman would be at an altitude of about 2,000 feet. By this calculation the relative uplift of the Sierra Nevada measured by the displacement of the base of the Penman Formation is about 5,000 feet.

In the description of the Penman Formation I noted that fossil plants indicate that the altitude of the summit region when they lived was not more than 2,000 feet. It is not possible to place the fossil plants closely with respect to the top and bottom of the formation as can be done with the Bonta Formation, so it is possible to say only that the indicated absolute uplift is also about 5,000 feet. The identity of these two values should be interpreted as coincidental rather than confirmational.

The Warner Formation cannot provide an absolute measure of uplift because it does not contain fossil plants, but, if one assumes that it is beneath the Mohawk Valley, as previously indicated, the relative uplift of the Sierra Nevada since the eruptive period of the Warner lavas was completed is more than 4,000 feet.

Neither does the Old Erosion Surface provide a measure of absolute uplift. Its relative uplift at Meadow Valley and at Big Hill in Plumas County is 1,500 and 2,000 feet, respectively.

Obviously the amount of uplift of the Sierra Nevada depends on the location and on whatever reference surface one uses to measure it. The variations indicate that it is not everywhere the same for any one reference surface. It seems likely also that the uplift was not continuous, but progressed in steps, with periods of rapid uplift alternating with periods of rest. Such variations would explain in part the characteristic profile of the east slope of the Sierra Nevada, which is convex upward, an example of which is shown in fig. 128, the slope from Haskell Peak to Mohawk Valley. Discontinuous uplift is also indicated by the nature of the Mohawk Lake Beds described farther on.

It is fair to conclude at this point that any answer to the question How much has the Sierra Nevada been uplifted? is not a sensible one unless one also asks, and can answer, the question During what period of time? That is true even for relative uplift. Much more information than is now available is needed to evaluate absolute uplift.

Should readers be unimpressed by the descriptions and numbers presented above, I urge them to visit some of the places described below to obtain a visual appreciation of the magnitude of the fault displacements that separate the Sierra Nevada and Grizzly Ridge from the narrow and relatively depressed Plumas Trench that lies between them. All are easily reached by automobile.

1. On the south side of State Highway 70-89 west of its junction with the county road to Sloat, the fields are underlain by lavas of the

Fig. 131. Mt. Jackson from the junction of State Highway 70-89 and Sloat Road. The upper dark part of Mt. Jackson consists of lava flows of the Warner Formation that are 350 feet thick. The rock of the foreground is the same lava of the Warner Formation. The difference in altitude of about 2,000 feet is the result of movement on the fault system along the southwest side of Grizzly Ridge.

Warner Formation. This is in Sec. 11, T. 24 N., R. 11 E. The altitude here is about 4,225 feet. The base of the Warner is at an unknown depth below the surface, perhaps only a few tens of feet, but not more than 350 feet. From this point one can look up to the southeast to Mt. Jackson and see there 2,000 feet higher, at an altitude of 6,250 feet, the base of the darker-colored lavas of the Warner resting on the gray rocks of the Bonta Formation, as shown in fig. 131. That 2,000 feet is a fair measure of the sum of the vertical components of movement of the several faults that intervene.

2. From the Mills Peak lookout the base of the Warner lavas that cap both Penman Peak and Mt. Jackson is visible. Its altitude on Mt. Jackson is about 6,250 feet, and on Penman Peak it is about 6,550 feet. The base of the Warner Formation that covers Big Hill is also visible from Mills Peak in the wall of the canyon of Jamison Creek near Two Rivers (fig. 132). Its altitude there is only 4,500 feet, 2,000 feet lower than on Penman Peak and 1,600 feet lower than on Mt. Jackson. On the near side of Jamison Creek Canyon the Warner lavas continue beneath Mohawk Valley to even lower levels, although this fact is not

a

Fig. 132. **a,** The view to the northwest across Mohawk Valley from Mills Peak. The skyline is the crest of Grizzly Ridge, altitude 7,000 to 7,500 feet. The conspicuous light-colored cliff at the left center is of Penman Formation in the west wall of the lower Jamison Creek Canyon. The continuation of the cliff to the right is of Penman Formation overlain by lava of the Warner Formation. The top of Big Hill above the cliff is composed of lavas of the Warner Formation. At the center of the picture but not visible is Two Rivers at the confluence of Jamison Creek and the Middle Fork of the Feather River. At the right is Mt. Jackson below the skyline but recognizable by its cap of dark lava of the Warner Formation, whose base there is at an altitude of about 6,250 feet. The altitude of the base of the Warner Formation in the wall of the Jamison Creek canyon at Two Rivers is 4,500 feet. The difference of 1,750 feet is the result of faulting along the base of Grizzly Ridge. The gently rolling hills that are the top of Big Hill are part of the Old Erosion Surface. **b,** Diagrammatic explanation of **a.**

visible from Mills Peak. Thus from Mills Peak one can see evidence of the upward displacement of Grizzly Ridge relative to Big Hill and Mohawk Valley.

A small patch of the Warner lava is present on the crest of the Sierra Nevada west of the west end of Gold Lake and south of Round Lake. It is not visible from the Mills Peak lookout tower, but can be

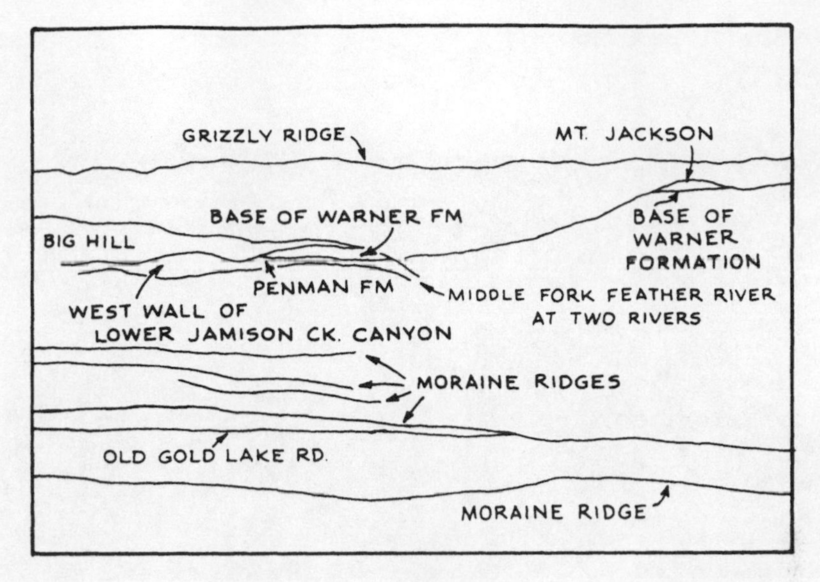

b

seen by walking a few tens of feet toward Gold Lake (to the southwest) from the parking area (fig. 133). It shows as a gray cliff on the skyline, and should not be confused with the black lava on the ridge on the northwest side of Gold Lake. The altitude of the base of this patch of Warner Formation is 7,500 feet. The difference of 3,000 feet between this point and Two Rivers is a rough measure of the uplift of the Sierra Nevada relative to Big Hill and Mohawk Valley.

3. From a point off the Eureka Ridge Road in the SW 1/4, Sec. 15, T. 22 N., R. 11 E., one can look northward directly down the lower canyon of Jamison Creek to Two Rivers on the Middle Fork of the Feather River. Beyond is Mt. Jackson. From here, as at Mills Peak, one can see the 1,600-foot difference in altitude between the base of the Warner Formation on Mt. Jackson and that in the walls of the Jamison Creek Canyon. This viewpoint, the altitude of which is 7,000 feet, is exactly at the base of the Penman Formation, which rests here on the Shoo Fly Formation. The Penman Formation is visible below on both sides of Jamison Creek, but its base is not exposed to view—it is below the level of the river at Two Rivers, altitude 4,500 feet. Thus, the uplift of the Sierra Nevada (Eureka Ridge) based on the Penman Formation, relative to the valley below, is more than 2,000 feet.

Although I have described these relationships as uplift of the Sierra Nevada and Grizzly Ridge relative to the valley between, I might as well have reversed this and written of the depression of the valley relative to the Sierra Nevada and Grizzly Ridge—that is what "relative" means.

Fig. 133. View to the southwest from Mills Peak. The low, dark cliff on the skyline to the right of the forest is the edge of a lava flow of the Warner Formation. The altitude of its base is 7,500 feet. The skyline is the crest of the Sierra Nevada west of Gold Lake and Round Lake. A part of Gold Lake is visible at the left edge of the picture, but Round Lake is not visible from this point. This view cannot be seen from the lookout tower, but can be had by walking a few tens of feet southwesterly from the parking area.

What I have described above as visible from each of several viewpoints is quite obvious, but the structural relationships are considerably more complex. The geologic cross section shown in fig. 129 gives some insight into the complexities; it shows some but not all of the relationships along a line from Eureka Ridge to Grizzly Ridge.

The essential facts are that on Eureka Ridge (Sierra Nevada) Penman Formation rests on metamorphic rock (Shoo Fly Formation). On Big Hill, Warner Formation rests on Penman Formation, which in turn rests on metamorphic rock. On Mt. Jackson, Warner and a thin layer of Penman Formation beneath it rest on the Bonta Formation, which is 1,000 feet thick and rests on metamorphic rock at an altitude of about 5,150 feet. There is no Bonta on Big Hill or on Eureka Ridge.

With respect to the Warner Formation, Big Hill (in the Plumas Trench) is depressed relative to both the Sierra Nevada and Grizzly Ridge. The same is true with respect to the Penman Formation. But consideration of the Bonta Formation shows that Grizzly Ridge is depressed relative to both Big Hill and the Sierra Nevada, and consideration of the basement of metamorphic rock indicates that Grizzly Ridge is elevated relative to Big Hill but depressed relative to the Sierra Nevada. Alas, these relationships cannot be accepted as a generalization about the three blocks, for they hold only along the line of the cross section in fig. 129. The relationships are different farther northwest and southeast, for the reason that the existing relationships are not just the result of events since the Warner Formation was erupted but are also the consequence of the earlier episodes of faulting and erosion mentioned at many places in Chapter 6.

CHANGES IN THE SYSTEM OF STREAMS

I provided some reasons in Chapter 6 for believing that until after the lavas of the Warner Formation were extruded, drainage from the interior across the site of the Sierra Nevada yet-to-be was uninterrupted. I believe that that condition existed until and during the time that the Old Erosion Surface was produced. No one has yet identified any main stream courses across that surface other than those that flow in the deep canyons, and they, except for the two main branches of the Feather River, head at the present divide.

The customary speculation has been that rapid uplift and tilting of the range caused an entirely new set of streams to come into existence, that they flowed directly down the newly tilted surface with increased slope, and that they headed at the divide because that was the eastern edge of the newly tilted slope. This simple story requires some modification.

The uplift and tilting was not a single, sudden, catastrophic event. It happened intermittently, and was the result of a series of events that occurred during a considerable amount of time. Such events continue even to this day.

Intermittent tilting must have produced intermittent changes in the then-existing streams. Erosion rates in the stream courses that flowed southwesterly would have been increased. Erosion rates on those that flowed northeasterly or southeasterly would at first have been but slightly affected. Streams that flowed north or east may have had their gradients reduced, or they might have been ponded or, in extreme circumstances, had their direction of flow reversed. There must have been a myriad of local changes in the stream system, the net result of which was the establishment eventually of a set of "new" streams that flowed more directly down the tilted Old Erosion Surface.

The "new" streams had increased erosive power as a direct result of tilting, which increased their gradients, and because their courses were shortened, which also increased their gradient. Thus the system of "new" streams grew out of the old by a process of reorganization. Gradually the "new" streams cut deep canyons into the Old Erosion Surface. Their tributaries also cut canyons, because tributaries, with rare exceptions, are constrained to meet main streams at grade. It is these "new" streams that occupy the present deep canyons; it was they that cut them.

Of course, factors other than those mentioned above were involved. Differences in the weatherability and erodability of different kinds of rocks are important. Local climate changes were no doubt important in causing an increased erosive power of the "new" streams, for the increased height of the rising mountains caused increased precipitation from passing storms. At the same time the region had begun to feel the effects of the general climate change that culminated in the accumulation of fields of snow and ice that fed the glaciers that were soon to appear. Such is the customary story, but it is not the correct or complete story.

Certainly much of the Old Erosion Surface was obliterated by the deepening and widening of the new canyons, and those parts that remain were, of course, also subject to continuous erosion. Perhaps one should think it remarkable that any of the present stream courses reflect their courses before the canyons were cut, but this is clearly so—including not only those streams upon remnants of the Old Erosion Surface between the canyons, but some of the canyon streams themselves. For example, the great loop of the North Fork of the Feather River around Big Bend Mountain was, no doubt, inherited from a similar if not identical loop upon the Old Erosion Surface. This is also true of the 8-mile-long, quite straight, southeast-trending course of the same river downstream from the loop (now under Oroville Reservoir) (see fig. 79) and the northwesterly course of Fall River.

The strange course of the Rubicon River (see fig. 79) probably is inherited also from the Old Erosion Surface. From its head southwest of the south end of Lake Tahoe it flows to the northwest for almost 20 miles, then makes a right angle bend and flows to the southwest. From there it follows a curved course and ends up trending again to the northwest. Thus, there is evidence from the main streams themselves that their courses across the Sierra Nevada are not entirely new—not the consequence only of southwestward tilting of the range.

I suggested earlier that before the uplift and tilting, streams on the Old Erosion Surface flowed across the Sierra Nevada. When uplift occurred they must either have maintained a course across the rising mountain front or have been beheaded or deflected. No streams flow around the Sierra Nevada so there is nothing to be said about deflection. The Middle and North forks of the Feather did maintain their courses and that matter is discussed further on. Consider the matter of beheading—the loss of the headward portion of a stream course.

A beheaded stream might be expected to occupy a valley that looks as though it ought to have a continuation into the air beyond the summit of the range, or to have a channel larger than is needed to carry the reduced flow caused by reduction of the area of the drainage basin. Although the landscape of the summit region has been greatly modified by glaciation it does appear that beheading can be detected in a few places with reasonable certainty.

The most obvious example is that of the South Fork of the American River, which heads at Echo Summit, elevation 7,377 feet, in an open valley at the very brink of the Sierran fault scarp, which is there 1,000 feet high. An eastward extension is easily imaginable, and it could have been through Luther Pass and the canyon of the West Fork of the Carson River.

Miller Creek, a tributary to the Rubicon River, heads in a broad pass west of Lake Tahoe at elevation 7,100 feet. The peaks adjacent to the pass rise many hundreds of feet higher. The pass appears to be in a valley whose eastward continuation now lies beneath the waters of Lake Tahoe.

The upper portion of the South Fork of the Yuba River has a quite gentle gradient and heads at Donner Pass at an elevation of 7,100 feet. Donner Pass is in a broad valley between Castle Peak 3.5 miles to the north, altitude 9,103 feet, and Mt. Lincoln, 2 miles south, altitude 8,383 feet. The valley occupied by Donner Lake, 1,100 feet below the pass, was strongly affected by glaciation, but it is not difficult to envision it as once a part of a headward course of the Yuba River.

A few miles farther north, Henness Pass Creek, a tributary of the Middle Fork of the Yuba River, heads at Henness Pass, a scarcely discernible divide in a meadow at an altitude of 6,900 feet. Nearby Webber Lake, which is, in effect, on the divide, is one of the sources of the eastward-flowing Little Truckee River. The area has been affected by glaciation. Faults are present, but the Sierran fault scarp only weakly developed is 5 miles farther east. If the land east of the crest were tilted eastward, the course of the Little Truckee River through Perazzo Meadows for a distance of 5 miles could have been a former course of the Middle Fork of the Yuba whose direction of flow has been reversed.

The North Fork of the Yuba River heads abruptly at Yuba Pass at an altitude of 6,708 feet, with peaks on both sides that rise 500 feet higher; the pass overlooks Sierra Valley 1,700 feet below. Although modified by glaciation, Yuba Pass appears to be only a little below the Old Erosion Surface. The probable former headward continuation of the river is now beneath Sierra Valley.

Eight miles northwest of Yuba Pass is the unnamed pass just east of Gold Lake. Its altitude is only 6,700 feet. Salmon Creek flows southward from Snag Lake to join the North Fork of the Yuba River, and Frazier Creek flows northward from Haven Lake and Gold Lake to join the Middle Fork of the Feather at Graeagle. This presents the intriguing possibility that Salmon Creek once extended farther north above the now strongly glaciated valley of Frazier Creek. Across Mohawk Valley lies Humbug Valley, extending from Clio to Portola. Humbug Valley is older than the arm of Mohawk Lake that occupied it, and it is not a fault valley. It seems to be of erosional origin on the Old Erosion Surface, and it may have been a segment of the North Fork of the Yuba system—that is, a headward extension of Salmon Creek cut off by the uplift of the Sierra Nevada (see fig. 130). Beyond Portola any continuation would now be beneath the northern part of Sierra Valley. This idea was proposed by H. W. Turner before 1900.

I am convinced that the streams mentioned above were beheaded by the uplift of the Sierra Nevada and relative depression of the region to the east. It should not be surprising that old westward-flowing courses on the eastern lands have not been recognized because that region has been greatly disturbed by faulting, by erosion, and by the creation of lakes and their filling with sediment. These matters are discussed further on.

If, as seems to be the case, the master streams of the western slope of the Sierra Nevada were beheaded, then it is erroneous to postulate that their courses were entirely consequent on uplift of the range—their present courses must be in large degree essentially as they were before uplift. That is to say that the present master streams of the Sierra Nevada follow nearly the same courses as did the master streams on the Old Erosion Surface. The entrenchment of the modern streams is the consequence of uplift, but their courses are not. Considerations of the courses of the Middle and North forks of the Feather River lead to the same conclusion.

The Old Erosion Surface of the Sierra Nevada is present on both sides of the canyon of the Middle Fork of the Feather River. On the north side it rises gradually from the Sacramento Valley to Clermont Hill, and on the south side it rises to Onion Valley. At both Clermont Hill and Onion Valley its northeastward continuation would extend out into the air, but it is recognizable as faulted down in the Plumas Trench, as previously described.

The crest of the uplifted Sierra Nevada across the canyon of the Middle Fork can be approximated by a line from Pilot Peak on Bunker Hill Ridge near Onion Valley to Clermont Hill. The junction of the river with Washington Creek at Minerva Bar is that point in the river almost directly under the projected crest. It is about 1 mile downstream from the junction with Bachs Creek, which marks the location of the most westerly fault of the fault zone that defines the eastern edge of the Sierra Nevada, and which is about 3,000 feet below the Old Erosion Surface.

Fig. 134. View to the west from Grizzly Ridge down the canyon of the Middle Fork of the Feather River west of Sloat. Long Valley is in the foreground. To the left is Eureka Ridge and to the right of center is Claremont Hill. The steep-sided notch in the skyline at the center of the picture is the Middle Fork Canyon between Little Volcano (Limestone Point) on the right and Quartz Point on the left. The canyon there is 3,000 feet deep. The viewpoint is in the SW 1/4, Sec. 1, T. 23 N., R. 11 E., above the forest road from State Highway 70-89 at Rattlesnake Creek to Happy Valley.

Obviously the canyon of the Middle Fork (figs. 134 and 129, section D–E) is incised into the Old Erosion Surface for its length of 40 miles from the junction of the river with the North Fork, now under Oroville Reservoir, to Minerva Bar. Why, then, does it not head there at a crest rather than some 11 miles farther upstream?

The head of the Middle Fork canyon is about 1.5 miles west of Sloat, between Sloat and the railroad tunnel to Spring Garden. From Sloat to Fells Flat, a distance of 5.5 miles, the river flows to the northwest, and at Fells Flat it is only 800 feet below the Old Erosion Surface, a figure that contrasts strongly with the 3,000 feet at Bachs Creek only 5 miles farther downstream. This westerly course of the river is controlled neither by a single fault, nor by a layer of soft and easily erodable rock. It seems to be a course inherited from the Old Erosion Surface.

From Fells Flat to Bachs Creek, a distance of 5 miles, the river crosses the eastern fault system of the Sierra Nevada, where the studies of Douglas Sheeks show that the faulting is so intense that the region is broken into fault-bounded blocks mostly less than a quarter of a square mile in area. Although the bed of the river along this course is in the metamorphic rocks of the Shoo Fly Formation, it is not far below the base of the Superjacent Series rocks, whose presence has permitted the recognition of the complexities of the faulting.

The traces of faults across the land are often marked by stream courses or ravines along them, or by saddles where the faults cross ridges. In other words, the courses of faults are often reflected in the details of the landscape. This presumably reflects the fact that rock adjacent to a fault is likely to be more intensely fractured than that farther away and is therefore more susceptible to erosion. Here, however, the course of the Middle Fork is not determined by any one or two faults, but by closely spaced faults that trend in many directions over an area of several square miles. The many faults evidently increased the erodability not just of a strip along a fault, but of a large volume of rock.

The answer to the question posed above, then, is that the course of the Middle Fork across the fault zone and the crestal region was established on the Old Erosion Surface before uplift occurred, and, perhaps fortuitously, it was located in an area of intense faulting and, therefore, easily erodable rock. For that reason it was able to maintain its course as uplift continued, cutting away its bed as rapidly as the Sierra Nevada block was uplifted. It was never beheaded. Thus the geological relationships along the Middle Fork confirm the conclusion that the course of the river across the Sierra Nevada was established on the Old Erosion Surface prior to uplift.

One should not be surprised at this explanation of the course of the Middle Fork, for rivers that have maintained a course across a rising mountain mass are numerous. Among others in California is the combined Sacramento and San Joaquin river system, which flows across the Coast Ranges by way of Carquinez Strait and the Golden Gate. Another notable example is the Russian River, which flows southward for 60 miles to the plain at the north end of the Santa Rosa Valley, from where it seems as though it might easily have made its way into the San Francisco Bay. Instead, it turns west and flows through a gorge across the Coast Range to enter the Pacific Ocean at Jenner.

The canyon of the North Fork of the Feather River, like that of the Middle Fork, also is incised into the Old Erosion Surface, which is preserved on both the north and south sides (see fig. 130). On the south, the Old Erosion Surface extends to the crest of the Sierra Nevada, marked by Spanish Peak and Mt. Pleasant just east of Bucks Lake. On the north, it extends to Spring Valley Mountain and an unnamed peak, altitude 7,088 feet, in Sec. 2, T. 25 N., R. 5 E.

The east slope of the Sierra Nevada south of the North Fork is the 2,000- to 3,000-foot-high scarp west of Meadow Valley. It is clearly indicated by the eastward course of Spanish Creek and its tributaries in the same way that the east slope along Mohawk Valley is marked by the northeastward courses of Mohawk, Frazier, Gray Eagle, and Jamison creeks. North of the North Fork there is no pronounced scarp, but the presence of the east slope is made evident by the east-flowing courses of Butt, Chips, Soda, Grizzly, and Yellow creeks.

The course of the missing crest of the Sierra Nevada across the North Fork canyon can be approximated by drawing a line from Mt. Pleasant to the unnamed peak on the north side. The altitude of the projected crest would be about 7,000 feet. The river directly below this projected crest about 1 mile west of Belden, at the mouth of Chips Creek, is at an altitude of about 2,200 feet, or 4,800 feet lower. This is, indeed, a very deep canyon, all cut during and since uplift.

The eastern fault zone, unlike that along the Middle Fork, is not easily identifiable because the rocks of the Superjacent Series have been lost to erosion. The North Fork crosses it between Canyon Dam and its junction with the East Branch, and the East Branch crosses it between the Y at the point of confluence of Indian and Spanish creeks and its junction with the North Fork near Belden.

The available geological studies of the region are concerned with the old metamorphic rocks, which do not lend themselves to an easy understanding of the much younger faults related to the uplift of the Sierra Nevada. However, the faults can be located approximately by the aligned stream segments and ravines eroded along them, by saddles where they cross ridges, by springs that are localized along them, and by the recognition of displaced areas of the Old Erosion Surface. On such evidence it is probable that no fewer than seven faults are crossed by the East Branch between the Y and its confluence with the North Fork.

The heads of the canyons of the tributaries of the North Fork are not immediately east of the fault zone, as is the case along the Middle Fork. Instead they are many miles farther east. On the North Fork proper, the head appears to be at Canyon Dam, 18 miles northeast of its junction with the East Branch, but it must really be much farther north, now concealed beneath the Almanor (formerly Big Meadows) basin. The canyon of the East Branch seems to head at Indian Valley about 6 miles upstream from the Y, but the streams headward of Indian Valley, which was once a lake, are also in deep canyons. The head of the canyon on Lights Creek is 21 miles upstream from the Y, and that of Indian Creek is 30 miles upstream from the Y. Little Grizzly, Wards, Red Clover, Crooks, and Moonlight creeks are also in deeply incised canyons.

The greater extent and depth of canyon cutting along the tributaries of the North Fork as compared to the Middle Fork seem to result from several factors. Together the North Fork and East Branch carry, on the average, more than three times as much water as the Middle Fork. This results in part from the fact that their combined drainage basin east of the fault system is more than twice as large as that of the Middle Fork. It seems also that the combined drainage basin of the North Fork and East Branch receives more precipitation than that of the Middle Fork, for rough calculations from measured stream flows show that the former yields more than half again as much water per square mile as does the latter. This, no doubt, is a result of the fact that the upper drainage basin of the Middle Fork, which is mostly Sierra Valley and its surroundings, is in the rain shadow of the Sierran crest, whereas that of the North Fork is more open to storms from the west.

It is my opinion that the foregoing description and discussion leads inevitably to the following conclusions:

1. It is obvious that the North Fork, like the Middle Fork, was not beheaded because it is continuous with tributaries that drain land 30 miles east of the Sierra Nevada (see fig. 130).
2. Incision of the North Fork into the Old Erosion Surface across the Sierra Nevada was accompanied by incision of its tributaries for many miles east of the Sierra Nevada.
3. The course of the North Fork and its tributaries, like that of the Middle Fork, was not determined by uplift of the Sierra Nevada, but was preexistent on the Old Erosion Surface.
4. Incision of the preexisting stream courses was the result of uplift and changes of climate related in part to the uplift, and in part to other causes.
5. Knowledge that the Middle and North forks of the Feather River were not beheaded by the uplift adds credibility to the notion that the main branches of the Yuba and American river systems were beheaded.
6. Knowledge that the courses of the Middle and North forks of the Feather River were determined on the Old Erosion Surface and not by uplift lends credibility to the notion that the courses of the main branches of the Yuba and American river systems also predate uplift.

THE EXTINCT LAKES

During the course of uplift of the Sierra Nevada, and the breakup of the region farther east, as shown in fig. 126, not only were the existing drainage lines interrupted, but closed basins developed on crustal blocks depressed relative to the blocks adjacent to them. Each closed basin became occupied by a lake.

In light of the length of geologic time, lakes are generally considered to be short-lived features of the landscape. The duration of a lake may be terminated in many ways, including filling of the basin with sediment, lowering of the river bed at the outlet by erosion, and evaporation of the water after a change of climate from wet to dry.

Lakes close to the Sierra Nevada, but now extinct, occupied Martis Valley, Sierra Valley, Mohawk Valley, Long Valley (at Sloat), Spring Garden, American Valley, Meadow Valley, Indian Valley, and the Almanor basin (called Big Meadows before Lake Almanor was created by damming the river). Lakes existed farther east in Little Last Chance Valley (now occupied by manmade Frenchman Lake), Grizzly Valley (now occupied by Lake Davis), Clover Valley and adjacent valleys, and Mountain Meadows, which also contains a reservoir (see fig. 38).

Other lakes appeared in Nevada, in the Great Basin, including giant Lake Lahontan, whose area was 8,600 square miles and whose maximum depth was 900 feet. Its last remnants are the alkaline Pyramid and Walker lakes and the now dry beds of Honey Lake and the Carson Sink. Lake Lahontan disappeared by evaporation. Of the many lakes that resulted from faulting in the region immediately east of the northern Sierra Nevada and in the Diamond Mountains only Lake Tahoe survives.

There is no reason to think that all the extinct lakes in northeastern California came into existence or disappeared at the same time. Neither is the creation of a lake basin an instantaneous affair. The large fault movements involved, like those of the uplift of the Sierra Nevada, accumulated during a span of time. Some lake basins filled with sediment and remain that way; some were drained before they became filled with sediment. Faulting and tilting may have occurred during the time of existence of a lake or after it became extinct. Thus each lake basin has its own distinct history. Unfortunately, the histories of the lakes in northeastern California remain largely unknown. In order to understand the history of a lake basin there should be either a considerable number of drill holes (wells), properly logged, or the sediments of the basin should be sufficiently eroded so that they can be studied on the surface. Drill holes are expensive, and those drilled for water are often not adequately logged. Of all of the lake basins in this area only three have been eroded: Meadow Valley, which has not been studied at all, and Little Last Chance and Mohawk, neither of which has been the object of thorough study.

It is convenient to have names for these extinct lakes. The name Lake Lahontan was established in the nineteenth century for the great

Fig. 135. Grizzly Valley looking south from a point near Bagley Pass toward the outlet by way of Big Grizzly Creek. Smith Peak is on the right and Crocker Mountain is on the left. This valley, formerly occupied by Grizzly Lake, is now partly occupied by manmade Lake Davis.

lake in western Nevada. In the last century also the name Mohawk Lake Beds was applied to the lake sediments that are present in Mohawk Valley. Used in this sense, Beds is synonymous with Formation. It seems appropriate to extend the name to the lake, hence Mohawk Lake. The names that I and my students have informally applied to other extinct lakes are noted below.

I have not described all of the lakes, only those that have some special interest. The order of presentation was chosen to develop concepts applicable to others that are described subsequently.

Grizzly Lake

Grizzly Lake is the name that I have applied to the lake that occupied Grizzly Valley (fig. 135). It was more than twice as large as artificial Lake Davis, which now occupies part of the same basin. Grizzly Lake became filled with sediment, leaving the quite flat surface of the valley, marred by only a slight amount of erosion near the present Grizzly Valley Dam. The floor of the basin is concealed from view, and there are no deep drill holes to aid in understanding its history. Nevertheless, because the geology of the surrounding area has been studied, it is possible to draw conclusions about the basin with confidence.

The basin is depressed along faults relative to the adjacent highlands. Figure 136 shows (in solid lines) the known and (in heavy dashed lines) the probable faults in the nearby region. The dominant faults trend northwesterly, and other faults trend in various directions between north and east. Those of the latter set are shorter than those in the northwest trend and tend to end against them. However, some faults in the northwest-trending set end against those of the other set. The approximate shoreline of Grizzly Lake, the 5,850-foot contour, is also shown on fig. 136 as a thin dashed line. The dotted lines represent possible faults close to or inside the shoreline that show how the basin might be bounded by faults parallel to those of the surrounding highlands. Although their positions and directions are inferred, they, with the known faults around the basin, make clear that the basin is complex; that it is not bounded simply by a few major faults in the dominant northwest trending set. Only in the manner shown can the curious crescent-shaped outline of the basin be accounted for. Were one to interpret the structure from the landscape alone one might conclude that the basin is bounded by a few strongly curved faults, but the absence of curved faults in the surrounding region precludes that possibility.

It must not be assumed from fig. 136 that faults are absent from the central part of the basin. On the contrary, it is quite probable that faults are as closely spaced and varied in direction as they are in the surrounding highlands, and that the basin is not the result of the depression of a single block, but of a series of adjacent blocks depressed to various amounts; it is a compound basin. In fact, the hill in the SW 1/4, Sec. 29, T. 24 N., R. 13 E., that stood as an island in the lake is the top of a block depressed less than blocks that surround it.

It is proper to think of the basin as the result of a local collapse of a part of the earth's crust. However, it must not be thought that this part of the crust simply fell into a great hole. The depression is accommodated by the extension associated with the normal faulting that characterizes the Basin and Range Province. This concept of the origin of the basin by a collapse or relative depression of adjacent blocks by differing amounts along normal faults is equally applicable to other basins of the region.

The depth of the basin is unknown, but no doubt it is hundreds of feet, possibly more than a thousand. As soon as any part of the basin became closed, it filled with water. The end product was Grizzly Lake, but there is reason, as is shown below, to think that faulting continued after the lake was fully established.

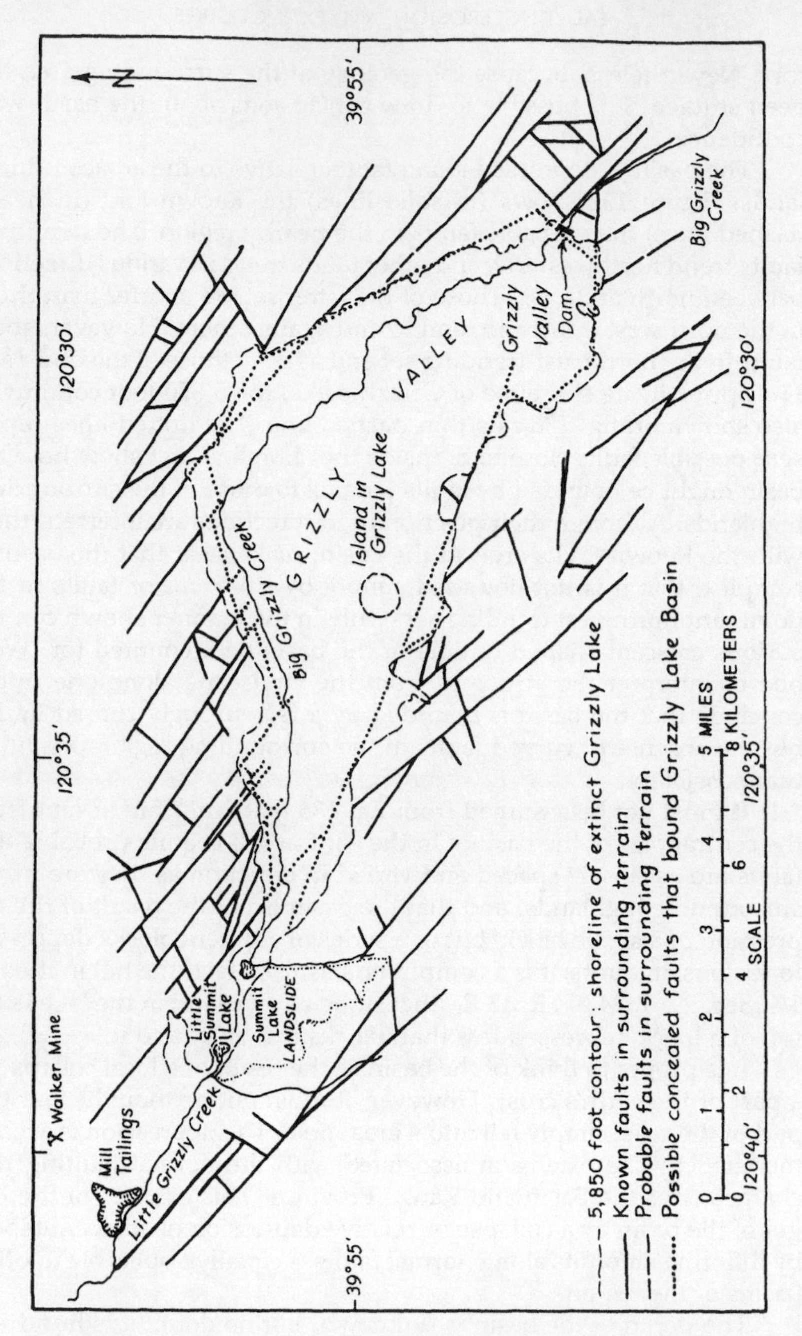

Fig. 136. A map showing the known faults around Grizzly Valley, and the possible positions of concealed faults that bound the basin.

Grizzly Valley is unusual in that it has two outlets that are within a few feet of being at the same altitude. Big Grizzly Creek, which drains virtually all of the surrounding mountains and the flat floor of the valley, leaves the valley at its southeast corner, where the present Grizzly Valley Dam is located. Below the dam the creek flows through a 6-mile-long canyon to join the Middle Fork of the Feather River 3 miles west of Beckwourth. In contrast, Little Grizzly Creek, which flows out of the northwest end of the valley, drains only a small part of the surrounding mountains and only a very small part of the valley floor. It flows through an 11-mile-long canyon to join Indian Creek in Genesee Valley. Thus it is part of the North Fork of the Feather River system.

The divide between the two streams is a swampy area in Secs. 20 and 21, T. 24 N., R. 12 E. Old maps show only one small lake, called Summit Lake, in the swampy area out of which Big Grizzly Creek flows. Later maps show correctly another called Little Summit Lake, out of which Little Grizzly Creek flows. The surface elevation of both lakes is close to 5,820 feet. No distinct channel connects the two lakes and they are separated by less than a mile of swampy land. The absence of a channel between them indicates that Grizzly Lake never overflowed down Little Grizzly Creek. But this raises problems concerning the altitude of the shoreline of Grizzly Lake.

Grizzly Lake was small—only 2.5 miles wide and 8 miles long. The prevailing southwest winds blew across the lake, not along it. Grizzly Ridge protected the lake from the wind, and so wave action was not strong; consequently, no pronounced shore terraces or beaches developed along the north and east shores. As a result, one can determine the position of the shoreline only by observing the altitude of the highest pebbles produced by waves at the shore, but these tend to be obscured or buried by rock and soil washed down from the slopes above. The island in Sec. 29, T. 24 N., R. 13 E. was protected from the latter effect because of its small size, so the information from there is probably the most reliable. On the island, the shoreline was at 5,825 feet, very close to the altitude of Summit Lake. On the north side of the lake near the Bagley Pass Road, pebbles are present as high as 5,875 feet, and on the east side they are present as high as 5,920 feet. Since, by natural causes, pebbles do not roll uphill, there is a discrepancy to be explained.

It is not possible to determine the altitude of the original spillover point near Grizzly Valley Dam by any independent means, but three-fourths of a mile southwest of the dam a saddle a few feet below 5,880 feet shows no evidence of overflow. This proves that the lake level was not that high, and certainly it was not as high as the pebbles at the 5,920-foot level on the east shore.

These facts lead to the conclusion that the shorelines on the east and north sides of the lake have been elevated by faulting. That tilting of the basin to the southwest did not occur is indicated by the fact that the valley floor slopes downward to the northeast and that the course of Big Grizzly Creek hugs the north and east margins of the valley (that part of the course now concealed by Lake Davis is shown on older maps). Neither does this northeast slope and the position of Big Grizzly Creek mean tilting in that direction; both are the result of the fact that most of the sediment that entered the basin came from Grizzly Ridge on the southwest side of the basin—that is where the longest and deepest canyons are.

It seems reasonable now to conclude that the original outlet by way of Big Grizzly Creek was close to 5,820 feet. However, the lake when first filled was in a delicate balance; it was within a few feet of spilling over to the northwest down Little Grizzly Creek, but the probability of that happening was reduced by either one or both of two events. There is a landslide south of the two Summit lakes. Its age relative to the lake is unknown, but it may have blocked any flow between the two lakes. However, a more probable cause is the fact that the outflow into Big Grizzly Creek was across deeply decayed and therefore easily eroded granitic rock. For that reason the maximum lake level might have been quickly lowered by as much as 50 feet once outflow began, thereby eliminating any possibility of overflow down Little Grizzly Creek.

Once the basin filled with sediment, sand and gravel could leave the basin by way of the outflowing stream and accelerate the lowering by abrasion of the stream bed at the outlet. When the water had been completely drained from the basin, a surface of soft, unconsolidated sediment was exposed, over which streams meandered and eventually joined together to form the upper course of Big Grizzly Creek.

The new stream joined the old lower course of Big Grizzly Creek at the outlet of the basin. This is also the place where the nearly zero gradient of the new stream on the newly exposed lake floor met the steeper gradient of the older stream course below. As this point was lowered by erosion, deepening of the channel on the lake floor became possible and the point of change of gradient then moved upstream; thus, the gradient of the new stream bed was steepened and its channel was deepened at the same time. This is the process called headward erosion, and all streams are capable of extending their courses headward. Anyone who has watched a gully develop on a hillside, even in so short a time as the duration of a single storm, will agree to the reality and effectiveness of the process.

Headward erosion by Big Grizzly Creek into the sediments of

Grizzly Lake had not progressed very far when the Grizzly Valley Dam was completed in 1966. At the dam site, which must be very close to the original point of outflow, the channel had been deepened by about 100 feet—that is the height of the dam. Deepening of the channel headward had extended about 2 miles upstream. The depth of erosion about a mile and a half north of the dam where a road formerly crossed Big Grizzly Creek was no more than 10 feet. The total amount of sediment removed from the basin was quite small, but enough to permit one to understand the process of headward erosion.

It was headward erosion into the Old Erosion Surface that permitted the eastward extension of the canyons of the streams tributary to the North Fork of the Feather River—those of Lights, Indian, Last Chance, Red Clover, Wards, and Little Grizzly creeks. Had headward erosion on Little Grizzly Creek progressed only a little farther, it, instead of Big Grizzly Creek, would have received the outflow from Grizzly Lake. Headward erosion certainly played a very significant role in the reorganization of the stream systems on the Old Erosion Surface when uplift of the Sierra Nevada occurred.

Indian Valley

Indian Valley has the highly irregular outline shown in fig. 137. The western portion, in which Greenville, Crescent Mills, and Taylorsville are situated, occupies a basin faulted down relative to the high mountains that surround it. Two armlike extensions of the western basin occupy Genesee Valley and the appropriately named North Arm. Both arms are in what were once deep canyons.

The faults that outline the western basin are not now visible—they are concealed by the lake sediments that fill the basin. However, the fault origin of the basin is indicated by the landscape and by the presence of a small patch of the Bonta Formation at the level of the valley floor in Secs. 35 and 36, T. 27 N., R. 9 E., a mile northeast of Greenville. The continuation of that bit of the Bonta Formation must have been above the surrounding ridges, no less than 2,000 feet higher on the northeast and 1,800 feet higher on the southwest. Faults in the surrounding mountains are obscure because the rocks of the Superjacent Series have been almost entirely removed by erosion, although some faults are reflected in the landscape. Not only do faults bound the basin, some also underlie it. It is a complex basin analogous to Grizzly Valley.

The shoreline of the lake is now marked by the edge of the lake sediments, which is essentially the same as the edge of the flat valley floor. In the western basin and in the North Arm, the altitude of the

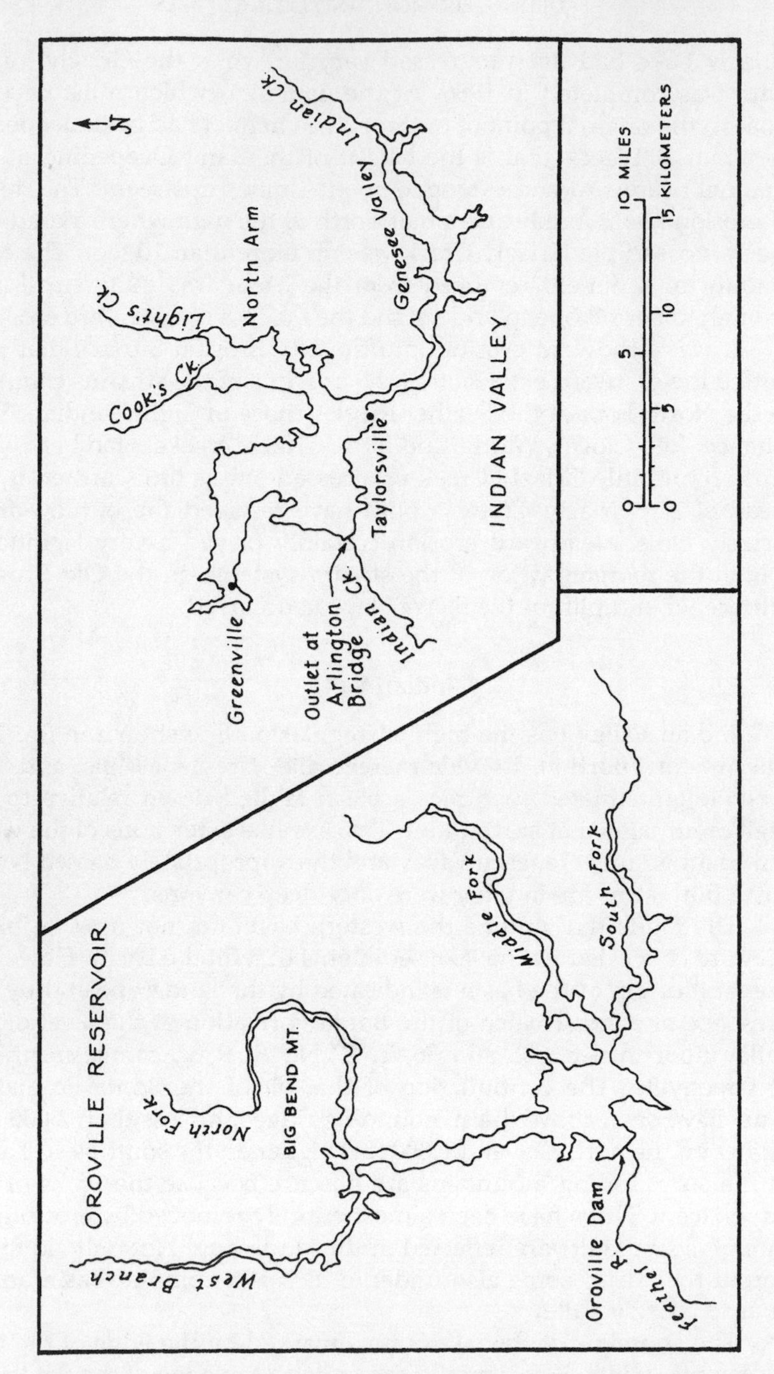

Fig. 137. Outline maps of Indian Valley and Oroville Reservoir at the same scale.

shoreline is about 3,600 feet; in Genesee Valley it is about 3,680 feet, which seems to indicate a small, late movement on a fault along Indian Creek southeast of Taylorsville.

The result of the major faulting was to depress an area that lay across the canyon of Indian Creek after it had been incised about 3,000 feet below the Old Erosion Surface. This, in effect, threw a dam across Indian Creek at Arlington Bridge, 1.25 miles southwest of Crescent Mills.

Whenever a stream is dammed, be it by man or by natural causes, the basin fills with water and overflows at the lowest point on the rim. The stream courses above the dam are submerged or drowned by the lake, which commonly leads to highly irregularly shaped lakes. It is interesting to compare the effect of the damming of Indian Creek by faulting and of the Feather River by manmade Oroville Dam. Both are on the same stream system and the resulting lakes are about the same size, Indian Valley being 23 square miles in area and Oroville Reservoir, 25. Oroville Reservoir has four narrow arms that extend up the tributary canyon. Indian Valley has only two arms but they have the same general form as those of the reservoir. The two basins are shown in fig. 137 at the same scale; their similarity is indeed striking.

When Oroville Reservoir becomes filled with sediment—gravel, sand, and mud—as it will in a few hundred years or less, it will be replaced by a meadow whose boundary will be that of the lakeshore, exactly like that of Indian Valley. The surface of the sedimentary fill will be almost as flat as the water surface of the reservoir. Indian Valley is that flat now and remains swampy in places despite the many drainage channels that have been constructed in it. This comparison should leave little doubt that Indian Valley is really a filled lake behind a dam even though nothing is known directly about the sedimentary fill.

The outlet from Indian Valley at Arlington Bridge, shown in fig. 138 from across the valley, is by way of the same canyon that existed before the faulting, and the stream bed is in hard metamorphic rock.

Lakes are traps for all the sediment that enters them. Because the erosion of a stream bed in hard rock is accomplished principally by the abrasive effect of rock particles—that is, sand grains, pebbles, and cobbles—being transported by the stream, the outflowing clear water from a lake cannot effectively lower the stream bed at the outlet. But once the lake basin is filled with sediment these "tools of abrasion" can cross the basin and again become active agents of erosion.

The filling of Indian Valley is so recent that the stream bed at Arlington Bridge has been lowered by only a few feet. Indian Creek has not yet extended its canyon into the basin fill by headward erosion, as has Big Grizzly Creek whose bed was in easily eroded, deeply decayed granitic rock.

Fig. 138. A view southwesterly across Indian Valley through its outlet at Arlington Bridge. The distant skyline is the crest of the Sierra Nevada in the vicinity of Mt. Pleasant.

The thickness of the sediment in Indian Valley, which is the same as the original depth of the lake, is unknown. However, projection of the gradient of Lights Creek from the head of North Arm to Arlington Bridge yields a rough estimate of 1,400 feet. This method can be tested at Oroville Reservoir. A projection of the gradient of the North Fork, 43 feet per mile between Belden and Pulga, under the reservoir to the dam yields a figure of 650 feet for the height of the dam. Of course the dam is actually 750 feet high, and the discrepancy is owing to the facts that the gradient of the river is not uniform, and the length of the channel below the reservoir cannot be accurately determined without resort to old maps. The estimated depth of the lake in Indian Valley is in error for the same reasons, and additionally because there are faults in the basin, which, because their sense and amount of motion is unknown, could make the estimated depth either too large or too small. In any event, it seems justifiable to conclude that the depth of the lake in Indian Valley was in hundreds of feet, not tens—it was a deep lake. It could well be that the volume of sediment under Indian Valley is equal to the volume of water in Oroville Reservoir.

Almanor Basin

Very little is known about the Almanor Basin and the extinct lake that occupied it. Although the area has not been studied, it is clear enough that the basin is the result of faulting that crossed and dammed the headwaters of the North Fork of the Feather River. The original lake surface was perhaps 150 feet higher than the present normal level of Lake Almanor, and the original outlet was probably about where it is now, over hard metamorphic rock. For this reason, lowering of the outlet must have been at first a slow process. However it is probable, although not certain, that the basin filled with sediment and that the North Fork canyon was deepened before the next notable event occurred, a series of volcanic eruptions from a source as yet unknown but one that was probably north of State Highway 36, a region of fairly recent volcanic activity.

Several lava flows poured down the North Fork Canyon, certainly as far as its confluence with the East Branch near Belden, and perhaps even farther. Lava filled the canyon near its head to a depth of several hundred feet. This event may have raised the lake level for a time, but the water soon overtopped the lava and again flowed down the canyon. The renewed stream flow has by now removed most of the lava from the canyon and has cut an additional depth of about 400 feet into the metamorphic rock below the lava.

American Valley

Very little can be said about American Valley and the lake that once occupied it because it has not been studied. However, some features are obvious. The lake resulted from the damming by faulting of Spanish Creek, in the same manner that the lakes in Indian Valley and the Almanor basin were formed. In this instance, however, it was the northeasterly flowing Spanish Creek, which heads on the crest of the Sierra Nevada, that was affected. The fault movement, relatively up on the northeast side, was on a prominent fault that lies at the base of Grizzly Ridge between Spring Garden and Keddie. Thus, originally the lake lay, and now the valley lies, in the Plumas Trench between the Sierra Nevada and Grizzly Ridge.

The basin is complexly faulted and must be composed of many fault-bounded blocks. The hills in the central part of the valley are the tops of high blocks that stood as islands in the lake. The outline of the basin, like that of Indian Valley, is that of a drowned stream system with arms that extend up Spanish, Greenhorn, and Thompson creeks. The recency of filling of the lake basin is indicated by the presence of swampy areas and of a pond in the springtime at the intersection of the La Porte Road and State Highway 70-89. Erosion at the outlet has progressed headward into the lake sediments by a distance of no more than a few thousand feet and to a depth of not more than 15 feet.

The altitude of the shore in the absence of a detailed study is uncertain. Examination of the best topographic map available indicates that in the Thompson Valley arm it may have been as high as 3,500 feet. Elsewhere it may have been as low as 3,440 feet. According to the map, if the lake level had been above 3,425 feet the lake would have overflowed at the saddle in Secs. 1, 2, and 35, 2.5 miles north of Quincy, where State Highway 70-89 leaves the valley, and spilled down Little Blackhawk Creek (fig. 139). That it did not do so is indicated by the absence of a channel through the saddle and the small size of the channel of Little Blackhawk Creek compared to that of Spanish Creek. In any event, the original outlet of the lake was where Spanish Creek now leaves the valley.

This discrepancy may be resolvable by detailed study. For example, it might be determined that faulting after the basin was filled with sediment has affected the altitude of the shoreline differently in different parts of the basin. On the other hand, the discrepancy may be only apparent, merely the result of attempting to read from the map detailed information it was never intended to convey.

The fact remains, however, that the basin had what may be thought of as a "potential outlet" by way of Little Blackhawk Creek, no more than a few feet higher than the actual outlet by way of Spanish Creek. Remembering that Grizzly Valley had such a "potential outlet" by way of Little Grizzly Creek and another by way of a saddle only 0.75 mile southwest of the actual outlet, it is apparent that such circumstances are not rare.

The importance of this discussion is evident in connection with the history of Mohawk Lake, related in a subsequent part of this report, wherein it is shown that the outlet to that basin was transferred by a process called capture from its original position to just such a "potential outlet."

Lake Ramelli

Frenchman Lake, about 9 miles north of the village of Chilcoot in Sierra Valley, is the result of the construction of Frenchman Dam, 110 feet high, across Little Last Chance Creek near the south end of Little Last Chance Valley. Little Last Chance Valley was previously occupied by a lake, informally called Lake Ramelli by John Van Couvering, who studied the region. Its shoreline is recognizable around the basin at an elevation of 5,800 feet, the highest level to which gravel is present (figs. 140 and 141). Most of the basin is still floored with sediment deposited in that lake.

The history of Lake Ramelli is somewhat better known than the history of those lakes already described because it was studied by a geologist, and because much of the sediment deposited in the lake

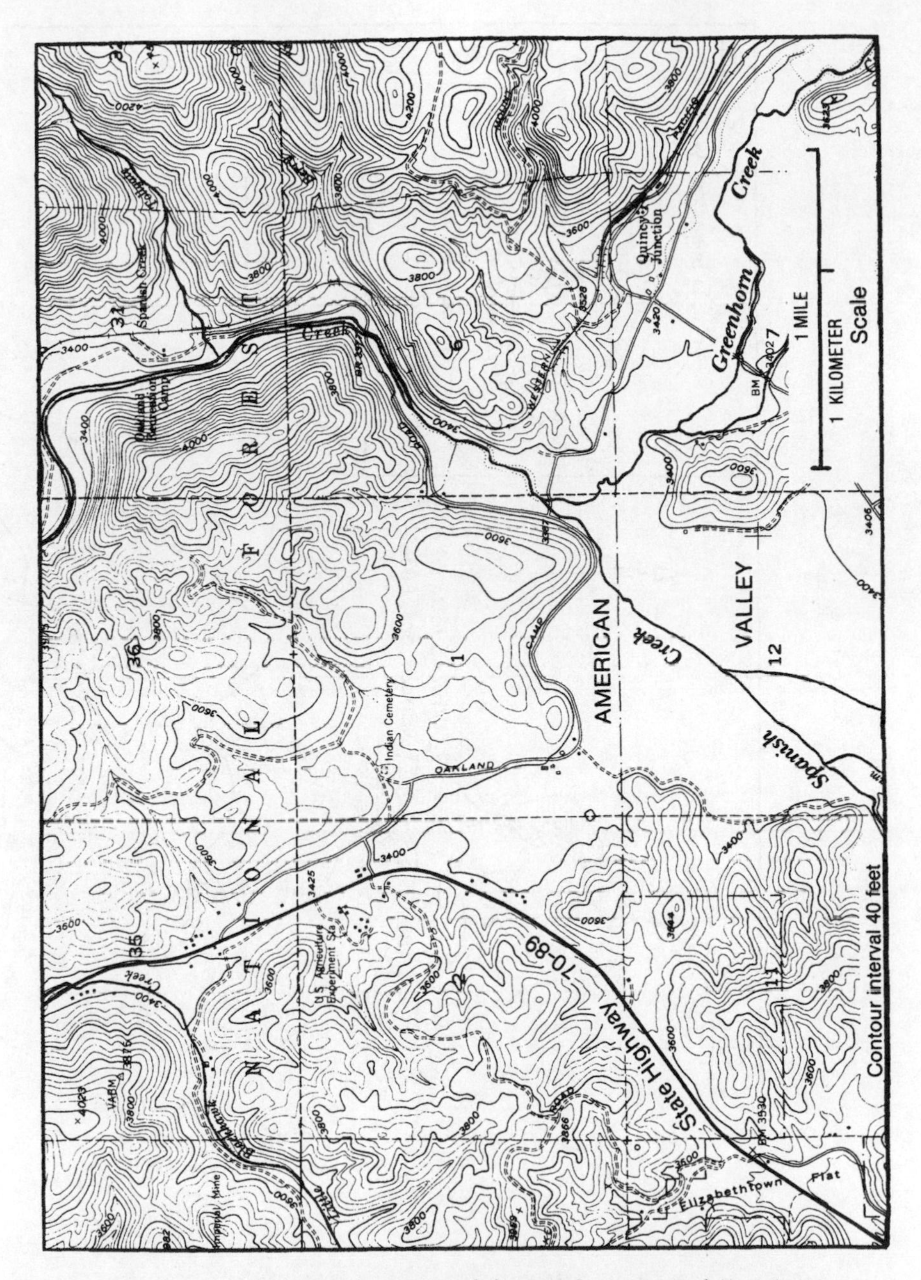

Fig. 139. A topographic map of the northern part of American Valley, showing its outlet by way of the canyon of Spanish Creek, and the shallow saddle at the northwest corner of the valley, in Secs. 1, 2, and 35, that leads to Little Blackhawk Creek.

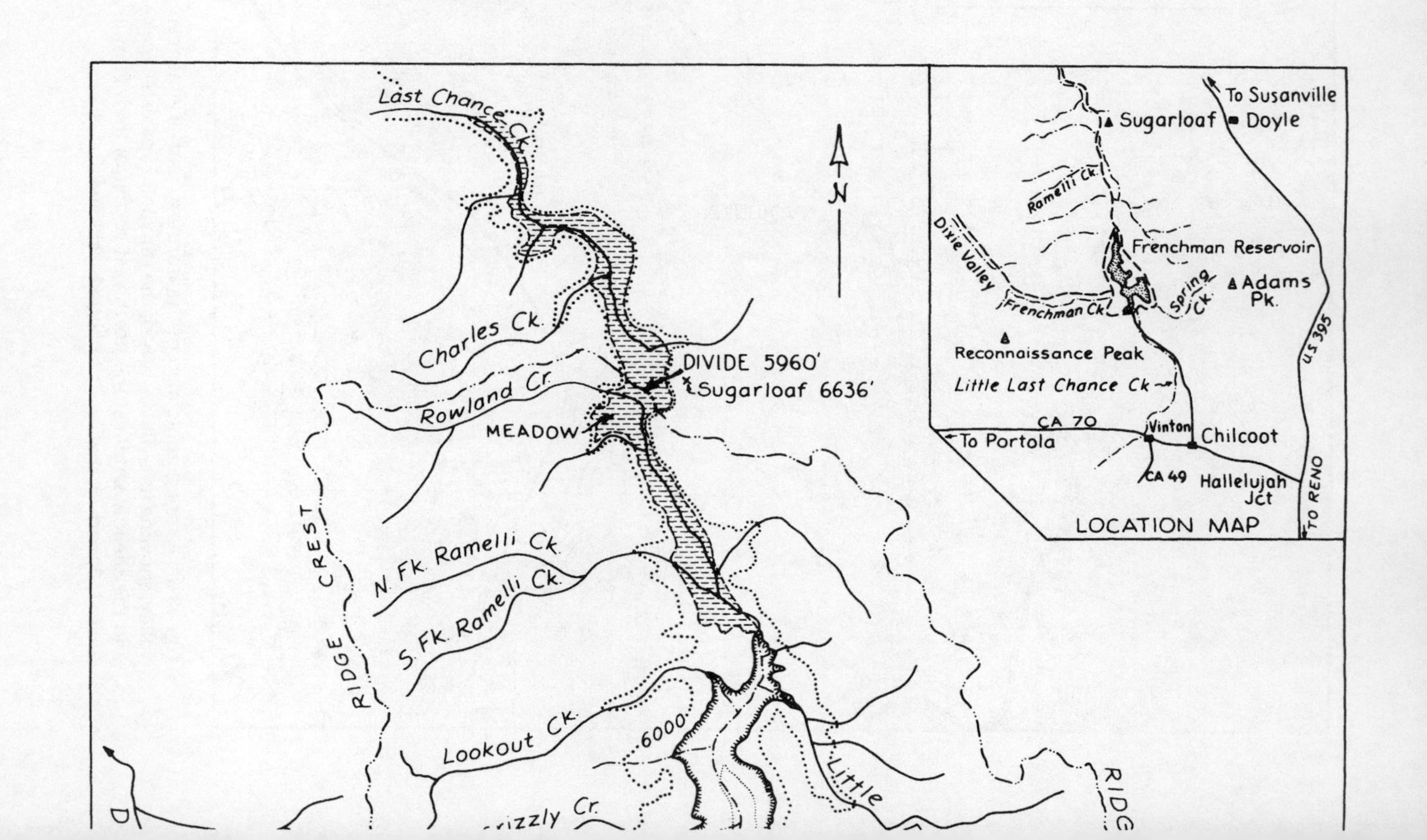

Last Chance Ck.
Charles Ck.
Rowland Cr.
DIVIDE 5960'
Sugarloaf 6636'
MEADOW
N. Fk. Ramelli Ck.
S. Fk. Ramelli Ck.
RIDGE CREST
Lookout Ck.
6000
Little
RIDG
N
To Susanville
Sugarloaf
Doyle
Ramelli Ck.
Dixie Valley
Frenchman Reservoir
Spring Ck.
Adams Pk.
Frenchman Ck.
Reconnaissance Peak
Little Last Chance Ck
US 395
CA 70
Vinton
Chilcoot
To Portola
CA 49
Hallelujah Jct
TO RENO
LOCATION MAP

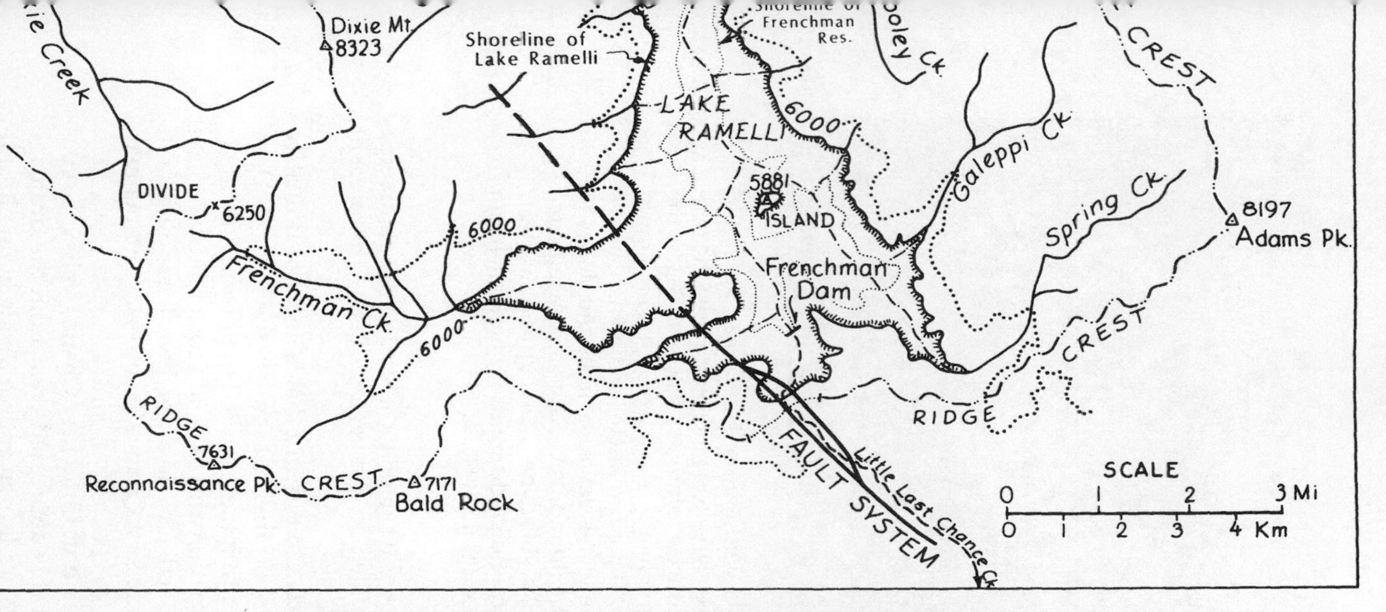

Fig. 140. Map of Little Last Chance Valley, formerly occupied by Lake Ramelli, and presently occupied by manmade Frenchman Reservoir. The heavy hachured line is the shoreline of Lake Ramelli at altitude 5,800 feet. The shoreline of Frenchman Reservoir at 5,590 feet and the 6,000-foot-contour are also shown. Note the position of the divide between Little Last Chance Creek and Last Chance Creek in a meadow at altitude 5,960 feet, shown also in fig. 142.

Fig. 141. A view along the west side of Frenchman Reservoir. The surface that shows as a nearly horizontal line across the center of the picture is the terrace at altitude 5,640 feet. That is the surface that resulted from a halt in the lowering by erosion of the outlet of Lake Ramelli. The more distant and higher surface at the center of the picture that is devoid of trees and slopes to the right is a remnant of the floor of Lake Ramelli. The left edge of that surface was the lake shoreline. Looking northwesterly from the road at the top of the grade south of the Frenchman Creek crossing. SW 1/4, Sec. 29, T. 24 N., R. 16 E.

had been removed by erosion. The geological study was completed before Frenchman Lake was filled so the geology below the present normal water level is known.

The basin is on a complexly faulted area that is depressed relative to a high ridge of granitic rock on the east, and to the ridge of Dixie Mountain on the west. The faulting that resulted in this relationship is younger than the Warner Formation and the Old Erosion Surface. The west boundary of the depressed area is a northwest-trending fault along which the canyon of Little Last Chance Creek below Frenchman Dam is incised.

Presently the drainage basin of Little Last Chance Valley is separated at its north end, within this depressed area, from the drainage basin of Last Chance Creek by an imperceptible divide at an altitude slightly below 5,950 feet in meadowland in Sec. 18, T. 25 N., R. 16 E. (fig. 142).

Fig. 142. The meadow in which the divide between Little Last Chance Creek and Last Chance Creek is located. Sec. 18, T. 25 N., R. 16 E.

Two lines of evidence suggest the possibility that Last Chance Creek once headed farther south. In the first place, alluvium in the valley of Last Chance Creek near the divide extends to an altitude of 6,000 feet, and is continuous with alluvium in the valley of Little Last Chance Creek at the same altitude. This suggests that a stream once flowed across the divide now occupied by meadow. The second bit of evidence comes from the way in which three tributaries of Little Last Chance Creek join the main stream.

The tributaries to main streams in hilly and mountainous regions generally join in a V that points downslope. If the V between joining streams points in the opposite direction, one should expect to find some special control other than the direction of the general slope. That condition is present here and is shown in fig. 140. Little Dooley Creek, tributary to south-flowing Little Last Chance Creek on its east side, flows to the northwest, and the upper courses of both Ramelli Creek and Lookout Creek on the west side flow to the northeast. The direction of Little Dooley Creek could have resulted from a fault along its course, but no such control is apparent for the other two. Alluvium in all three extends above the 6,000-foot level, so it is possible that they were once tributaries to a north-flowing stream that could be none other than Last Chance Creek. In the southern part of the basin the evidence is less clear, but the valley of Spring Creek contains alluvium to the 6,000-foot level, and that of Frenchman Creek to the 6,400-foot level. Neither of these streams trends northerly, but at the same time neither is clearly directed downslope into Little Last Chance Creek. These facts are not conclusive but they indicate the possibility that the entire drainage basin of south-flowing Little Last Chance Creek was once the headwaters of north-flowing Last Chance Creek.

Consideration of the gradient of Last Chance Creek helps one to understand what may have happened. Its average gradient for 27 miles north of the divide that separates it from Little Last Chance Creek is only 20 feet per mile. By projecting that gradient above Little Last Chance Creek as far as a mile south of the present dam, one finds that the altitude of the stream bed there would have been about 6,130 feet. This is the lowest possible height of a divide that would have separated such a former course of Last Chance Creek from Sierra Valley. There is a dividing ridge there now, except that it is breached by the sharply incised canyon of Little Last Chance Creek below Frenchman Dam. However, at a distance of less than a mile on both sides of the canyon, the ridge is higher than 6,000 feet. Thus it is possible that Last Chance Creek did once head at a former divide about 1 mile south of Frenchman Dam.

The conversion of a gently northward sloping valley to the closed basin of Lake Ramelli can be accounted for by a renewed movement on the fault that bounded the originally depressed area on the southwest side, which is also the fault now followed by Little Last Chance Creek in its canyon below the present dam. The required movement would have been a further depression of about 700 feet accompanied by westward tilting of the depressed block. This would have provided a basin whose original depth of 300 feet would have been adequate to contain Lake Ramelli. It would also have brought a saddle, or pass, whose altitude before faulting was 6,200 feet, down to 5,800 feet, the level of Lake Ramelli, thus providing a southern outlet from the newly created basin.

Very likely there was already a stream flowing down the south side of the divide into Sierra Valley, probably along the fault where Little Last Chance Creek is today below Frenchman Dam. The added overflow from the newly created Lake Ramelli would have augmented its volume and its erosive power, but its rate of downcutting was perhaps not greatly increased until the lake basin filled with sediment.

The sediment deposited in Lake Ramelli that still remains is mostly sand and gravel, but finer-grained sediment was deposited in the deeper central part of the lake. A hill in the SW 1/4, Sec. 21, T. 24 N., R. 16 E., whose summit altitude is 5,887 feet, was a small island near the center of the extinct lake, but at present it is a peninsula extending out from the east side of Frenchman Lake. Traces of gravel remaining on this hill indicate that the lake basin did, indeed, fill with sediment.

Once the lake basin was filled with sediment, and sand and gravel began to be carried down Little Last Chance Creek, downcutting of the stream bed could have been quite rapid, for the course of the stream is along the complexly faulted and jointed rock along the fault. As the bed of the outflowing stream was lowered, the sedimentary fill of the basin was gradually removed.

After a time there was a halt in the erosion, as indicated by the presence of another terrace and beach, or bench, around the basin at an altitude of about 5,640 feet (see fig. 141). This event was probably the result of the exposure in the bed of Little Last Chance Creek of a body of harder rock, more resistant to erosion. Several of these bodies are present along the stream course below Frenchman Dam. Once this obstacle was removed by erosion, the stream could again continue to lower its bed until Frenchman Dam was completed in 1961.

The history of this basin is thus quite different from those postulated for Almanor, Indian Valley, and American Valley. They were formed by fault-caused dams across main streams. Here faulting was the apparent cause, but tilting of the depressed fault block probably also occurred, and the direction of the slope of the valley was reversed. This is, of course, a speculative history; the true history may never be known. Other histories may be devised by imaginative geologists, and I have thought up others, but they seem less probable and do not explain as many of the observed facts as does this one.

Lake Ramelli flowed into Lake Beckwourth in Sierra Valley, but there is no evidence that relates the two lakes in time with any degree of precision.

Lake Beckwourth and Sierra Valley

Lake Beckwourth, informally named by John Van Couvering in honor of the pioneer James Beckwourth, occupied Sierra Valley. Its shoreline, at an altitude of about 5,100 feet, is not easily seen around most of the basin although it is marked by gravelly terraces that are really old beaches. However, an impressive series of shorelines at various altitudes below 5,100 feet is visible in afternoon light beside State Highway 49 on a hill 4 miles north of Loyalton, and on another hill 5.5 miles west of Loyalton (fig. 143). The shoreline is also fairly well marked on the hill slope southwest of the town of Beckwourth and on the east side of the Buttes, 3.5 miles east of Beckwourth, which once

Fig. 143. The shorelines of Lake Beckwourth in Sierra Valley appear as horizontal lines across the hillside. The uppermost one marks the highest stand of the lake surface at an altitude of 5,100 feet. The lower lines are former beaches, or strands, that are evidently the result of intermittent lowering as a consequence of erosion of the outlet of the lake near Beckwourth. Looking east from State Highway 49 at the hill in Sec. 17, T. 21 N., R. 15 E., 5.5 miles west of Loyalton.

stood as an island in the lake. The area of the lake was about 180 square miles. The outlet was by way of the canyon now followed by the Middle Fork of the Feather River west of Beckwourth.

Excluding a small part of the valley immediately west of Beckwourth, the floor of Sierra Valley is the exceedingly flat surface of the sediment deposited in the lake. Its altitude near Beckwourth is 4,810 feet; it rises to about 4,900 feet only close to the shoreline. A large part of the surface remains swampy in spite of many manmade drainage channels. The maximum depth of water at the time drainage of the lake began was close to 230 feet.

The basin occupied by Lake Beckwourth originated by faulting; faulting relatively downward with respect to the surrounding lands, and to a few blocks that remained high to become islands like the Buttes. Although the geology of the surrounding country is well known only along the north and northwest sides, it is apparent that the basin had an origin like that of Grizzly Valley; the principal faults trend northwesterly and lesser faults trend between north and east. One could postulate a set of faults that are the possible or even probable boundaries of the basin, as I have done for Grizzly Valley in fig. 136. It is the Sierra Nevada frontal fault system that bounds the basin along its extreme southwestern margin. There the east front of the Sierra Nevada is the steep rugged slope more than 2,100 feet high that extends southward from Sattley, and that is so striking as one drives westward from Sierraville on State Highway 89.

Not much is known about the depth of the basin, but one would expect that below the lake sediments there are the same kinds of rocks in the same kinds of relationships as are present in the surrounding highlands. Indeed, a well near the center of the basin drilled in search of geothermal energy penetrated 1,275 feet of lake sediment below which there are 950 feet of volcanic rock of the Superjacent Series. Below that the drill encountered granitic rock. Adding the depth of water in the lake to the thickness of the lake sediments shows that the original depth of the basin at the well site was about 1,500 feet. Other wells indicate that the lake sediments throughout the central part of the basin are probably no less than 1,000 feet thick. It may be that in some places the lake was even deeper than 1,500 feet. This value compares rather well with the present maximum depth of Lake Tahoe of 1,600 feet.

Although, as remarked above, the outlet of the lake was by way of the existing canyon of the Middle Fork between Beckwourth and Portola, it is tempting to think that the lake may once have overflowed at Beckwourth Pass into the Great Basin (fig. 144). That, however, did not happen. The altitude of the pass is 5,218 feet, more than 100 feet higher than the lakeshore. Furthermore, the rock at the pass is easily erodable, deeply decayed granitic rock; if water had flowed across the pass there would be a deep gorge leading down into Long Valley, and at least a part of Sierra Valley would now drain that way.

The only possible outlet for the water of the basin is the canyon between Beckwourth and Portola. Very likely this canyon, then a more open valley, was, at an even earlier time, a principal drainage channel for much, if not all, of the area tributary to Lake Beckwourth. The distribution of gravel that marks the shoreline of Lake Beckwourth around the east end of the canyon indicates that the overflow occurred at an altitude close to 5,100 feet—the highest lake level.

The water that flowed westward through the channel entered Mohawk Lake at Portola. There the shoreline of Mohawk Lake, close to 5,040 feet, was 60 feet lower than that of Lake Beckwourth, so the short river that connected the two was probably a fast-flowing stream. From the time of first overflow until the present, the bed of that river has been lowered to an altitude of 4,870 feet, 23 feet below the original

Fig. 144. Beckwourth Pass. Looking east from State Highway 70 between Vinton and Chilcoot. The pass is the low, open saddle at the center of the picture. The distant skyline is that of Petersen Mountain in Nevada.

level. While this erosion occurred, Lake Beckwourth was gradually drained.

When the level of Lake Beckwourth was reduced to that of Mohawk Lake, at an altitude of 5,040 feet, lowering of the lake should have stopped and a new shoreline should have been produced at that elevation, an event that seems to have happened. The many shorelines noted above as visible on hill slopes both north and west of Loyalton (see fig. 143), although not accurately surveyed, appear to lie between 5,100 feet, the highest lake level, and the 5,040-foot-level of Mohawk Lake at Portola. They reflect stages in the lowering by erosion of the bed of the outlet channel between Beckwourth and Portola.

When the level of the two lakes became nearly the same, Lake Beckwourth was still about 160 feet deep, and its surface could not be further lowered until Mohawk Lake was drained. That seems to have been a quite sudden event, but whether there are shorelines around Lake Beckwourth below the 5,040-foot level that reflect further lowering of the lake is not yet known.

Since Mohawk Lake was drained, headward erosion of the Middle Fork has cut into the sediments of Lake Beckwourth to a depth of only about 10 feet, as seen at the bridge on the county road south of Beckwourth, and has extended headward no more than a mile upstream from the bridge.

Although the shorelines of Lake Beckwourth have not been studied in detail, no obvious evidence has yet been found that would indicate that the basin has been either tilted or faulted since the lakeshore beaches were formed.

Mohawk Lake and the Mohawk Lake Beds

The name Mohawk Lake Beds was applied in 1891 to the young sedimentary rocks in Mohawk and Humbug valleys. It seems appropriate to use the same name for the lake in which the sediments were deposited.

Mohawk Lake was T shaped with the head of the T, which I call the Mohawk arm of the lake, about 14 miles long and 4 miles wide, occupying Mohawk Valley. The stem, 8 miles long and 2 miles wide, occupied Humbug Valley between Clio and Portola (fig. 145). I call that the Humbug arm. Remnants of the shoreline and considerable areas of the lake floor, well preserved all around the basin, show that the altitude of the lake surface was between 5,000 and 5,050 feet, most probably 5,040 feet. The outlet of the lake was at its most westerly point, by way of Chris Creek and Poplar Creek, thence into Long Valley at Sloat. The evidence for this is presented further on. The history of the lake ends when it was drained; at that time the depth of water near Graeagle was about 175 feet. Since that time, a thickness of almost 500 feet, measured vertically, of the sediment deposited in the lake has been removed by erosion. For that reason it is possible to understand some of the early history of the lake, which is not possible for any of the other nearby lake basins.

The two parts of the basin had different origins and different histories. The bottom of the lake sediments in Humbug Valley is well exposed in many places. Although the surface on which the sediments rest is irregular, it has the general altitude of about 4,880 feet, so the Humbug-arm lake had an original water depth of only about 150 feet.

As I indicated in my discussion of the uplift of the Sierra Nevada, it appears that Humbug Valley was not of fault origin but was erosional, and that possibly it was eroded by a stream that was then tributary to the North Fork of the Yuba River. If that was its origin, it was cut off by the uplift of the Sierra Nevada and relative depression of the land to the east.

In contrast, the Mohawk arm is certainly of fault origin. It is a deep basin depressed between two roughly parallel sets of faults—those of the east front of the Sierra Nevada that bound it along Mohawk scarp and Eureka Ridge, and those of the set along the southwest side of Grizzly Ridge. The total depth of the basin is unknown, but the bottom of a 400-foot-deep well in the Mohawk Valley is in fine-

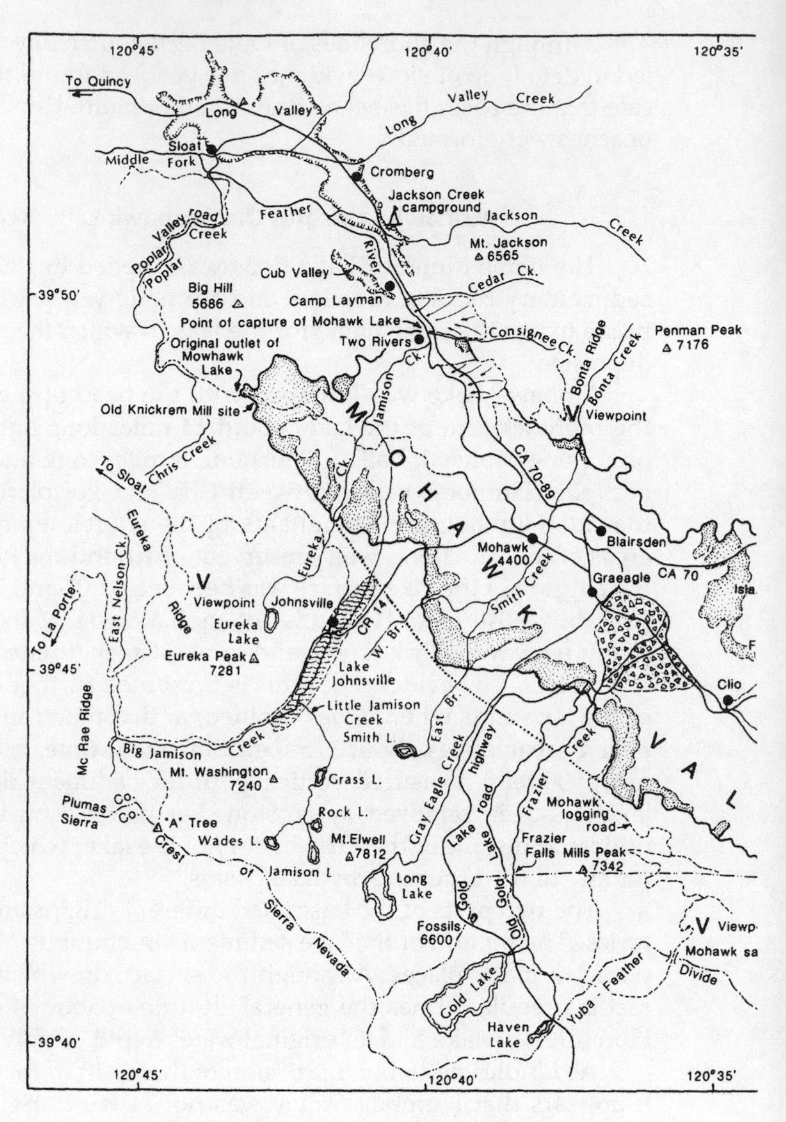

Fig. 145. Map of Mohawk Lake and vicinity showing the remaining portions of the lake floor.

grained lake sediments at an altitude of 3,900 feet. This indicates a minimum depth of 1,100 feet. However, some geophysical evidence indicates that near Clio the remaining lake sediments are 1,500 feet thick, or that the base is at an altitude of only 2,900 feet, and that the basin was originally more than 2,000 feet deep.

It was made known in the last century by H. W. Turner that the sedimentary rocks of the central part of the Mohawk Valley were fine

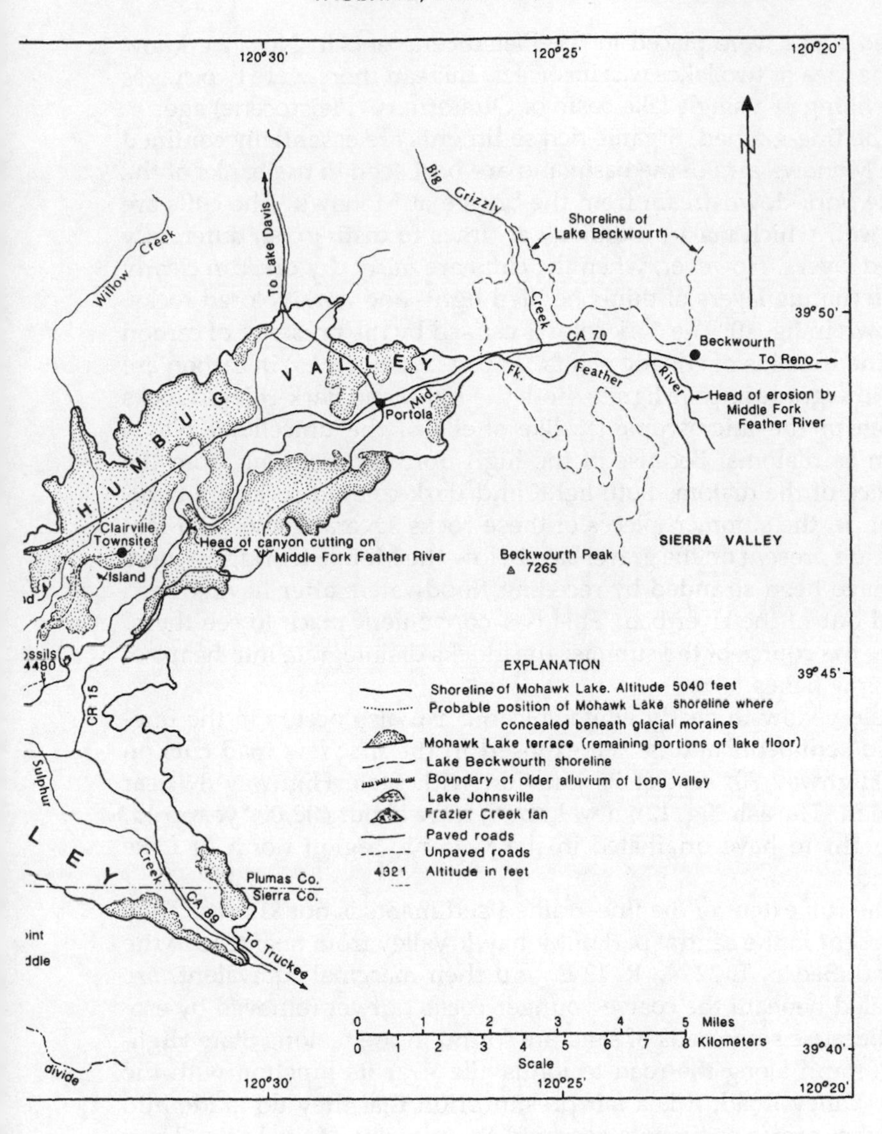

grained and rich in organic matter, and that they were overlain by sand and gravel that contain no recognizable organic remains. This led to the idea that two lakes of different ages had successively occupied the same basin. The mistaken belief that the fossil-leaf-bearing clays near Clio, now known to be in the Bonta Formation, were part of the fine-grained Mohawk Lake Beds resulted in assigning the "older," fine-grained beds to the Miocene epoch. The "younger," coarse-grained rocks were placed in the Pleistocene epoch. Now we know that the idea of two lakes was incorrect. Instead there were two stages in the filling of a single lake basin of Quaternary (Pleistocene) age.

The fine-grained, organic-rich sediments are essentially confined to the Mohawk arm of the basin and are best seen in the banks of the Middle Fork downstream from the bridge at Mohawk. The cliffs are often wet, which makes it difficult at times to distinguish differently colored layers. However, when the cliffs are fairly dry one can clearly see alternating layers of thinly bedded light- and dark-colored rocks, as shown in fig. 10. The dark color is caused by the presence of carbon from the remains of aquatic plants. Those layers richest in carbon are really low-grade impure lignite. Both the light- and dark-colored rocks are rich in the microscopic lacelike shells of the unicellular plants known as diatoms. Because of the high porosity resulting from the presence of the diatoms both light- and dark-colored rocks are lightweight. In the summer, pieces of these rocks several inches to a foot across are present on the gravel bars below the Mohawk Bridge, where they have been stranded by receding floodwaters after having been ripped out of the river bed. This is a convenient place to see them. During the course of the summer the blocks disintegrate into heaps of small gray flakes.

Pale yellow to nearly white volcanic ash also occurs in the fine-grained sediments and is well exposed in the first few road cuts on State Highway 70, east of its junction with State Highway 89 near Blairsden. The ash (fig. 12), now known to be about 400,000 years old, is thought to have originated in the volcanic region north of Lake Almanor.

The full extent of the fine-grained sediments is not known. They are present in the central part of Mohawk Valley from near Clio to the center of Sec. 5, T. 22 N., R. 12 E., but their marginal equivalents are concealed beneath the coarse younger rocks not yet removed by erosion. Because some beds of sand are found in them along State Highway 70, and along the road to Johnsville near its junction with the Poplar Valley Road, it is a fair presumption that they do grade into somewhat coarse sediments nearer to the margins of the basin. Their thickness is also undetermined, but they are present at the bottom of the 400-foot-deep well mentioned above, and their top seems to be not much above 4,400 feet, which indicates a thickness of not less than 600 feet. Of course, what is below the bottom of that well remains unknown. Pollen grains from the layers rich in carbon show that the basin was surrounded by a forest of the same coniferous trees that grow there today, which means that the altitude and the prevailing temperature in and around the basin were much as they are today.

The conclusion to be drawn from this meager amount of information is that the lake represented now by the sediments visible at the level of 4,300 to 4,400 feet was one with shoaling shores along which aquatic plants grew in profusion; that the waters were well fertilized and the free-floating microscopic organisms represented by the diatoms flourished; that the bottom waters may have been stagnant—that is, depleted of oxygen—which would account for the preservation of so much of the organic tissue as carbon; that whereas sand entered the basin from time to time, the principal sediment was mud, organic matter, and volcanic ash transported by air from a remote source. The absence of gravel indicates that the lake was not closely surrounded by high mountains as the valley is today. Perhaps at that time the lake was not very deep, but it persisted because subsidence relative to surrounding highlands kept pace with filling of the basin, which, since it was largely organic, was at a slow rate.

After the deposition of the fine-grained sediment as high as the present altitude of about 4,400 feet, a sudden change occurred; from then on, sand and gravel became the dominant sediments deposited in the lake. This event may mark the onset of glaciation, or a rapid uplift of the Sierra Nevada to near its present height, or, most likely, a combination of both. The older, finer sediments were eroded; channels cut in them were backfilled with gravel, and pieces of the diatomaceous and carbonaceous sediment occur in the gravel. Even perfectly rounded pebbles of the fragile carbon-rich sediment have been found in a layer of coarse sand in the ravine just below the junction of the Johnsville Road with the Poplar Valley Road (SE 1/4, Sec. 8, T. 22 N., R. 12 E.).

The sediments were no longer deposited in quiet waters; rapid deposition and turbulence is indicated by cross beds and channels eroded and then filled by the subsequent layer. The sediments are poorly sorted; that is, mud, sand, and pebbles are mixed together. Organic remains have not been found.

On the south side of the lake, from which most of the sediment came, Jamison and Gray Eagle creeks, and probably Frazier Creek as well, built fan-shaped deltas into the lake. On the northeast side, the fewer and smaller streams from Grizzly Ridge dumped their load of sediment over the steep fault scarp; in the ravines one can see that lake sediments rich in boulders grade rapidly into sand outward from the buried face of the ridge.

Meanwhile, the shallower Humbug arm of the lake filled with sediment brought in by Willow and Humbug creeks and other smaller streams (fig. 146). The coarse gravels of this part of the lake are visible along State Highway 70 and the old highway (formerly U.S. 40A) between Blairsden and Portola and along the unpaved road between Clio and Claireville townsite. The very coarse gravel shown in fig. 147 is in

Fig. 146. Mohawk Lake Beds. Cross-bedded sandstone and conglomerate in lenticular beds with cut-and-filled channels, features typical of deposition from fast-moving currents. These sediments were probably deposited in water about 450 feet deep at the junction of the shallow Humbug arm with the deeper Mohawk arm of the lake. On the road from Clio to Claireville townsite.

the first road cut on State Highway 70 west of the bridge across Willow Creek.

The oldest glacial sediments found thus far in the region are in the Mohawk Lake Beds along the north branch of Smith Creek (recently named Claim Creek) in the SE 1/4, Sec. 17, T. 22 N., R. 12 E., a mile southwest of Mohawk. Here, at an altitude of about 4,600 feet, a layer of poorly sorted sediment contains blocks of rock as much as 4 feet long that have the flattened and striated surfaces that are the characteristic results of transportation by glaciers.

Like the lakes already described, Mohawk Lake can be seen in the landscape. A view across the valley from an altitude of 5,000 feet or more in almost any direction will include some flat areas close to the 5,000-foot level that are remnants of the lake floor or terrace. The largest of them, crossed by State Highway 70 between Willow Creek and Delleker, is only slightly touched by erosion. One of the best and easiest places to visit the lake floor is at Claireville townsite. Figure 148 shows this remnant and a hill that stood as an island in the lake.

Fig. 147. Mohawk Lake Beds. Conglomerate of large boulders deposited in shallow water. The flat surface above the road cut is the Mohawk Lake terrace, the floor of the lake. Note the figure at the right of the center for scale. The first road cut on State Highway 70 west of the bridge across Willow Creek, between Blairsden and Portola. SE 1/4, Sec. 7, T. 22 N., R. 13 E.

The nearly flat area crossed by the road from Mohawk to Johnsville at the boundary of the Plumas-Eureka State Park is part of the terrace. Although not very wide at the road crossing, the terrace continues both ways for a considerable distance. Madora Lake, the result of the damming of a shallow ravine, is on this part of the terrace. One can see the shoreline on the face of the mountain on the northeast side of Mohawk Valley, as shown in fig. 149, and the road that crosses the face of the mountain passes along some small remaining flat areas at the very shoreline. Gold Lake Highway crosses the lakeshore just west (upgrade) from its junction with Old Gold Lake Road in Sec. 28, T. 22 N., R. 12 E., a mile south of Graeagle. Farther upgrade (toward Gold Lake), the flat below the right (north) side of the road is part of the lake floor. The steep slopes on the left (south) are on glacial deposits.

The most spectacular view of all is from the Bonta Ridge Road on Grizzly Ridge in the NW 1/4, Sec. 4, T. 22 N., R. 12 E. (fig. 150). Looking across the Mohawk Valley one sees the east front of the Sierra

Fig. 148. The Mohawk Lake floor or terrace at Claireville townsite. The hill in the background is a plug of Warner basalt that stood as an island in the lake. Secs. 8 and 18, T. 22 N., R. 13 E.

Nevada from Mt. Haskell on the southeast to White Cap on Eureka Ridge on the northwest. In the center are the long glacial moraines parallel to Frazier, Gray Eagle, and Jamison creeks that extend northward from Mills Peak, Mt. Elwell, and Eureka Peak. In front of the tips of the great moraine ridges is the Mohawk Lake terrace, on which the moraines rest. The northern tip of the terrace, a mile southwest of the Graeagle business center, is close to the center of the lake. Its altitude of about 4,866 feet indicates that the Mohawk arm of the lake was not entirely filled with sediment before it was drained; above Graeagle, water remained to a depth of about 175 feet.

The original outlet of Mohawk Lake is reached by way of the Poplar Valley Road from its junction with the Johnsville Road a mile west of Mohawk. Beyond Squirrel Creek the road climbs to the flat, gravel-strewn lake floor, and a mile farther on is the Old Knickrem Mill site, the location of a sawmill, long since disappeared, where several forest roads come together. Here in an open meadow in the SW 1/4, Sec. 35, T. 23 N., R. 11 E., is the beginning of the outlet channel from the lake. Farther northwest the channel is a flat-floored, graveled surface a quarter mile long with mountain slopes rising on both sides. The downward gradient of the channel as far as the west line of Sec.

Fig. 149. The edge of the Mohawk Lake terrace on Grizzly Ridge as seen from the Mohawk Cemetery. The terrace shows as a nearly horizontal line midway across the view at the top of a series of small open spaces in the forest. Its ends are indicated by the black arrows. The shoreline was probably a few tens of feet higher. The forest road from the Bonta Ridge Road to the Denton Creek Road follows along the terrace remnants.

35 is barely perceptible. One must keep in mind that all of this surface has been repeatedly disturbed by logging operations and road construction, but this quarter-mile-long stretch is the low point—the spillover point—on the rim of Mohawk Lake. Beyond the west line of Sec. 35 (fig. 151), the gradient of the channel increases, and near the middle of Sec. 34 there is a short, narrow gorge between two cliffs of the hard metamorphic rock of the Taylor Formation (fig. 152). The gorge has also been extensively modified by road construction since I first saw it in 1939. At that time it was a deep slot between two towering masses of rock. This must have been a spectacular cascade when the outflow of Mohawk Lake, to which all the upper Middle Fork basin was tributary, passed through.

Presently, from the Knickrem Mill site to a point below the gorge, a distance of a mile, there is no permanent stream in this old river channel. Just below the gorge, tiny Chris Creek enters from the slope of Eureka Ridge and trickles along through a canyon it could never

a

Fig. 150. **a,** The Mohawk Lake terrace south of Gray Eagle in Sec. 21, T. 22 N., R. 12 E., as seen from the Bonta Ridge Road across the valley. The viewpoint is marked on figs. 145 and 158. The ridge that rests on the lake terrace is the moraine on the west side of Gray Eagle Creek. The line of trees along the base of the moraine ridge marks the approximate position of the shore of Mohawk Lake. **b,** Diagrammatic explanation of **a.**

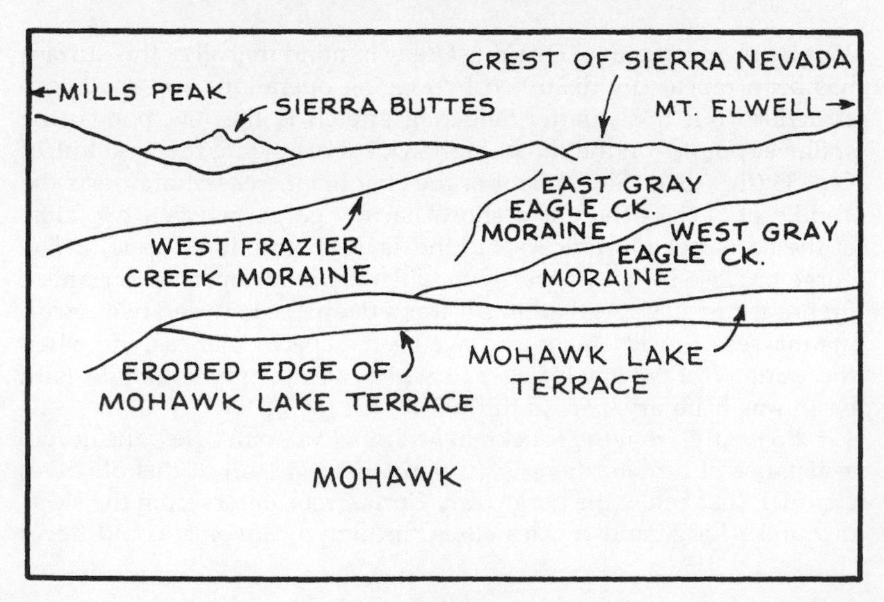

b

Fig. 151. The outlet of Mohawk Lake at the west line of Sec. 35, T. 23 N., R. 11 E., along the Poplar Valley Road. The ground is much disturbed by repeated road grading and logging.

Fig. 152. The gorge along the now dry channel of the original outlet of Mohawk Lake. The river was probably not less than 40 feet deep at this point. The figure of a man standing in front of the cliff near the center of the picture sets the scale. A quarter mile west of the site of fig. 151, in the SE 1/4, Sec. 34, T. 23 N., R. 11 E., along the Poplar Valley Road.

Fig. 153. The valley of Poplar Creek 5 miles downstream from the outlet of Mohawk Lake, and 1.5 miles south of Sloat in the NW 1/4, Sec. 22, T. 23 N., R. 11 E. This valley carried the entire flow of the Middle Fork of the Feather River before Mohawk Lake was drained. The result of the drainage of the lake was to leave this valley much wider than is necessary to carry the flow of the much smaller Poplar Creek, and to permit it to become choked with alluvium from the valley walls that the present stream, because of its small size, is unable to carry away.

have cut. Downstream the water from large springs localized along the faults that limit the east front of the Sierra Nevada join to form Poplar Creek, which, after a course of about 6 miles, enters Long Valley and joins the present Middle Fork of the Feather River at Sloat. The entire course of Poplar Creek is alluviated, in many places with obvious terraces on one or both sides (fig. 153). It gives the distinct impression that it flows in a valley wider than is needed to carry the present flow. That is true—its valley did once carry the flow of the Middle Fork, then a very much larger stream than it is now.

The Untimely End of Mohawk Lake

As I have already indicated, the life-span of Mohawk Lake did not come to an end because it filled with sediment. It met its end by the establishment of a new outlet at a different place. This process, called capture, was not involved in any of the other lakes described.

In my description of the canyon of the Middle Fork of the Feather River below Long Valley, I pointed out that the head of canyon cutting was 1.5 miles west of Sloat. Of course, Long Valley was tributary to the Middle Fork, and Poplar Creek, carrying the outflow from Mohawk Lake, was the main stream.

Long Valley is a small down-faulted basin with many faults in and around it. Possibly it once held a lake, although that is not proven. However, most of its floor is covered with alluviums of two ages. Both consist of coarse gravel that came from the surrounding canyons, especially those of McDermott Ravine, Little Long Valley Creek, Long Valley Creek, Cogswell Ravine, and Jackson Creek. The younger alluvium is related to the present course of the Middle Fork, but the older one, as shown below, is older than the drainage of Mohawk Lake. One can travel on the surface of the older alluvium almost without interruption from Sloat past Cromberg to the Jackson Creek Campground, where State Highway 70-89 leaves the alluvial surface and begins its passage along the wall of the canyon of the Middle Fork, which leads upstream into Mohawk Valley. At the Jackson Creek Campground, the highway crosses the edge of the alluvium, as shown in fig. 154. The altitude here is 4,450 feet, 240 feet above the river and 330 feet higher than the river at Sloat. The slope of the alluvial surface, 97 feet per mile, is much steeper than that of the river, 71 feet per mile. Looking downriver, one can see the edge of the alluvium exposed in the cliff above the railroad track sloping downward toward Sloat (fig. 155).

From this point a narrow strip of the same alluvium continues another 4,000 feet southeasterly along the canyon wall and attains an altitude of about 4,720 feet; beyond that it has been eroded away. A matching strip of alluvium, not apparent from the highway, is present on the south side of the river also. It ends above the south end of the bridge to Camp Layman, also at an altitude of about 4,720 feet. All of these relationships are shown on the map of the Long Valley area (fig. 156).

It is obvious that the Long Valley alluvium is older than this course of the river because the canyon in which it flows is cut through it. Jackson Creek does not flow out of its canyon at the campground directly into the Middle Fork; it maintains its old course on the alluvial surface for another mile. It then joins the Middle Fork near the Cromberg Cemetery. Cedar Creek, like Jackson Creek, almost certainly once flowed over the alluvium and contributed to it, and it is possible that Consignee Creek did also in spite of the fact that it now flows

Fig. 154. The older alluvium of Long Valley and its sloping surface near the entrance to the Jackson Creek Campground. The view is of a road cut on State Highway 70-89 at the northwest end of the canyon of the Middle Fork between Two Rivers and Cromberg. The alluvial sand and gravel are rust colored, in contrast to the light-colored mudflow breccia of the Penman Formation on which it rests.

directly into the Middle Fork. Although erosion and two large and much younger landslides on the north side of the river—one in Secs. 29 and 30, the other in Sec. 32, of T. 23 N., R. 12 E.—have obliterated most of the evidence that would have revealed the details of the history of this area, the general picture is reasonably clear.

It is significant that a bit of the Mohawk Lake shoreline at 5,000 feet altitude, and the terrace with sediment about 50 feet thick, is present on the ridge on the east side of Consignee Creek. This ridge is between the two landslides but it is stable and was not affected by either of them. The gap between this relict of the Mohawk Lake terrace and shoreline and the eroded end of the alluvium of Long Valley is 6,000 feet, a little more than a mile. Projection of the Long Valley alluvial surface to the southeast, parallel to the river, shows that it would pass above this remnant of lakeshore and that it cannot, therefore, have been connected with the lake; there must have been a divide between the two. The altitude of the divide, which would have

Fig. 155. The older alluvium of Long Valley. A view to the northwest into Long Valley and down the Middle Fork of the Feather River toward Cromberg from a point above State Highway 70-89, 0.5 mile southeast of the Jackson Creek Campground. The surface of the old alluvium slopes away from the camera, and the alluvial gravel is exposed along the edge of the slope above and on the right side of the river and railroad. Almost this same view can be had from the highway.

been about over Two Rivers, cannot be estimated accurately; it may have been more than 100 feet higher than the lake level, or it may have been as little as 10 or 20 feet above it. The situation is reminiscent of the low divide between the lake that occupied American Valley and Blackhawk Creek in Sec. 2, T. 24 N., R. 9 E., which was only a few feet too high to have become the outlet.

How then was the outlet of the lake by way of Poplar Creek along the southwest side of Big Hill transferred almost 3 miles to the north to flow along the northeast side of Big Hill? Two facts are important in providing an explanation: (1) The rock in the walls and bottom of the new canyon is relatively soft and easily eroded; it is mostly the mudflow breccia of the Penman Formation. (2) The canyon is located along the zone of faulting at the base of Grizzly Ridge. Figure 156 shows that the basalt lavas of the Warner Formation are faulted down relative to Mt. Jackson. The base is dropped from 6,200 feet to 4,760 feet on the northeast side of the river at Two Rivers, and to 4,500 feet on the

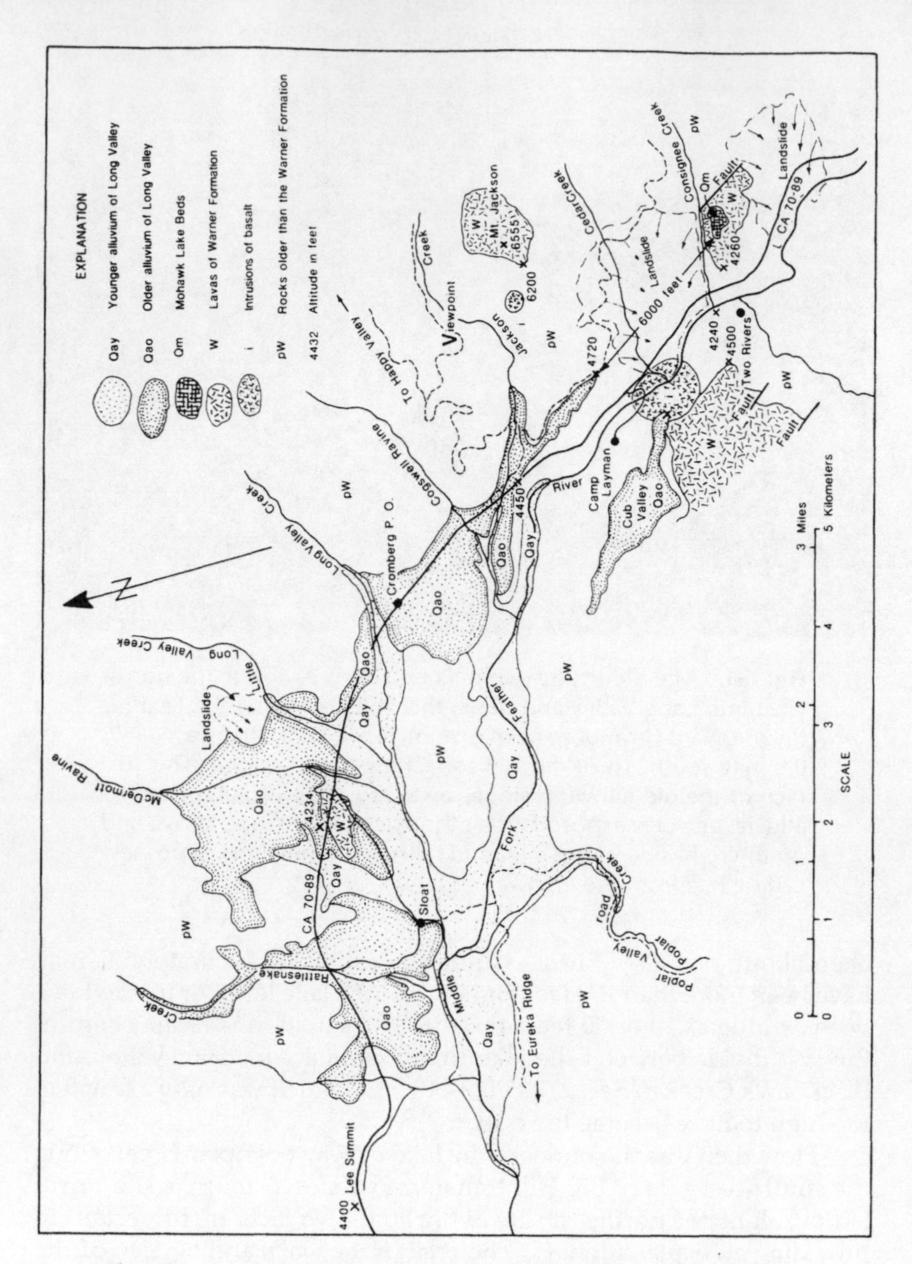

Fig. 156. A partial geologic map of Long Valley and of the canyon of the Middle Fork of the Feather River leading upstream into Mohawk Valley to show the relationships pertinent to understanding the capture and drainage of Mohawk Lake. See also fig. 145.

southwest side. Both the original nature of the rock and the faulting made the rock along this new course easily erodable once an overflow was established.

Of course, the divide was subject to erosion as a consequence of flowing water from melting snow and rain, but the lowering of the divide, sloping gently to both sides and covered with vegetation, would have been a very slow process. However, another agent of erosion—winddriven waves on the lake—was also at work. The prevailing wind direction is from southwest to northeast, but occasionally, especially in summer, tropical storms bring powerful winds from the southeast along the length of Mohawk Valley. Such winds blowing along the lake for nearly 12 miles would not only produce large waves at the northwest end but would for short periods of time raise the lake level, which would accentuate the erosive attack on the divide. Figure 157, for example, shows how winddriven waves cut into the east shore of Lake Davis from 1966 to 1977 to make a bench and a cliff more than 8 feet high. Here at Lake Davis the rocks attacked are the soft sediments of Grizzly Lake, but the wind fetch is less than 2 miles.

After a sufficient number of storms had passed, the divide would no doubt have been lowered to such a level that storm waves began to splash over it. This may have happened many times before a permanent outflow from the lake was established, but once that occurred the trickle soon became a torrent. The rock was soft, and the slope into Long Valley was steep; the amount of water available was large, for, as has already been shown, Lake Beckwourth as well as Mohawk Lake was involved. A minor catastrophe was under way—a new outlet to the lake was established.

That is my preferred interpretation of the events leading to the drainage of Mohawk Lake. There is, however, one other, perhaps equally probable history. Mohawk Lake existed through most or all of the Ice Age, the subject of the next chapter, when the climate was colder than it is at present. It is possible that the lake froze in winter. An ice dam at the outlet near the Old Knickrem Mill site, such as could have resulted from the driving by wind of floe ice into that end of the lake, could have blocked the outlet and permitted the lake level to rise by several tens of feet. Once the divide between the lake and the Long Valley alluvial surface had been lowered sufficiently by erosion, a single such ice dam could have caused the initial overflow, and possibly only one such event would have been sufficient to permanently establish the new outlet.

Whatever the cause of the overflow, the consequences of the establishment of a new outlet were the same. At first, as the channel deepened, more and more water poured through it, and soon sand

Fig. 157. The east shore of Lake Davis at the Grasshopper Flat Campground. The cliff, about 8 feet high, was cut by waves at times when the lake level was high during the eleven-year interval 1966–1977.

and gravel of the lake sediments became available to scour the channel bottom. As the lake level was lowered, the courses of the Middle Fork and all its tributaries that had entered the lake flowed across the surface of the unconsolidated sediment of the newly exposed lake floor sediment that those same streams had brought into the basin. Headward erosion of the main channel upstream from the place of the no longer existing divide rapidly removed the lake sediment. Each tributary was constrained to deepen its channel to meet the Middle Fork at grade. Each stream thus shared a part in excavating the basin to its present form, amounting to almost 500 feet vertically and 1 to 3 miles horizontally. The process of removal of the lake sediment continues today, and only in very recent time did headward erosion in the Middle Fork reach Portola. The head of erosion is now about 1 mile upstream from the bridge at Beckwourth in Sierra Valley.

Removal of the Mohawk Lake Beds has, in places, revealed the rocks below them, especially in the Humbug arm, where the Middle Fork has cut a new canyon upstream from Clio. This has not happened in the deep part of the Mohawk arm, although rock below the lake beds is now exposed in places near the margins in such a way as to reveal the fact that the once concealed valley walls are steeply sloping fault scarps. For example, the Penman Formation is visible along Smith Creek 1.5 miles southwest of Mohawk, and the Bonta Formation is faulted down against the Peale Formation near the center of Sec. 4, T. 22 N., R. 12 E., about 1 mile north of Mohawk Ranger Station.

The interesting course of the lower part of Jamison Creek was determined by the configuration of the surface of its own deposits in the lake. The deposits comprised a fan-shaped delta, probably arched in the center and lower on the sides. When the lake level was lowered, the newly extended course of the stream was along the northwest side of the fan close to the lakeshore to join the Middle Fork directly above the place now called Two Rivers.

Jamison Creek was joined then as now by Eureka Creek in the extreme northeast corner of Sec. 1, T. 22 N., R. 11 E., only a short distance downstream from the crossings of both creeks by the Poplar Valley Road. On cutting downward through perhaps no more than 100 feet of lake beds, both streams encountered rock of the Penman Formation. The Penman was no obstacle to downward erosion but below the Penman, which is there no more than 150 feet thick, is the hard rock of the Sierra Buttes Formation. This area of the older rock, probably no more than a half mile across, had stood high on the original basin floor, but not as an island—it had been completely buried by lake sediment. Once the streams had cut down to the Sierra Buttes Formation they were confined to their new courses, and, amply provided with sand and gravel, "the tools of abrasion," they continued to cut away even this resistant material, making the present almost vertically walled gorge, about 160 feet deep, within which the two streams join. Thus, the present courses of these two streams and the place of their junction were superimposed upon the hard rock of the Sierra Buttes Formation by their initial courses, established in a different body of rock, under different circumstances, at a higher level.

That the drainage of Mohawk Valley was a geologically recent event is evident from the fact that the story is inscribed in the landscape. But it is difficult to establish a date for it.

From the glacial history, described in Chapter 8, it is quite clear that the lake existed as recently as 60,000 to 75,000 years ago during the Tahoe glaciation, and that it probably existed as recently as 20,000 years ago when the Tioga glaciation reached its maximum extent. Another approach yields a still younger date, as is shown in Chapter 8.

8

Origin of the Landscape

II. THE AGE OF GLACIERS

INTRODUCTION

A large part of the magnificent scenery of the mountainous regions of the world is the result of sculpturing in stone on a grand scale by glaciers—masses of ice in motion. The great Ice Age of the Quaternary period that began 2 to 3 million years ago was a worldwide event of tremendous scope and importance. The Laurentide Ice Sheet of eastern North America alone covered more than 5 million square miles, much of it more than a mile thick. The volume of water thus withdrawn from the sea and stored on land as ice was enough to lower sea level by 250 feet. The lowering of sea level changed radically the geography of the edges of the continents and had profound influences on life both in the sea and on land. Furthermore, this was not a single event—four times during the Quaternary period this huge mass of ice developed and then melted. Each of the four glaciations was also complex and involved several advances and retreats of the ice. Similar ice sheets appeared and disappeared synchronously in Europe.

Enough ice remains today in Greenland and Antarctica to raise sea level by about 150 feet should it all melt. The fact that these two great ice sheets still exist, and that together they cover about 10 percent of the earth's land surface, may be taken to indicate that the Ice Age may not yet be over—we may well be living in a fourth interglacial, the name given to the periods of warmer and drier climate between the glacial episodes.

Beyond the limits of the great ice sheets of continental extent, ice also accumulated at higher altitudes as ice streams in the valleys of mountainous regions. It is these, called valley glaciers, with which we are concerned. They developed along nearly the entire length of the Sierra Nevada, on nearby high mountains east of the Sierra Nevada such as the Carson Range and Warner Mountains, in the Klamath Mountains, and on Mt. Shasta. Valley glaciers, like the Laurentide Ice Sheet, also appeared and disappeared several times. Glaciers remain on Mt. Shasta, and some very small glaciers are present in the high southern Sierra Nevada. A few small bits of ice called glacierets also remain, and a bit of ice that persists in a crevice on the north side of Sierra Buttes seems to be one of them.

I have not discussed the causes of glaciation because they remain unknown. It is well established, however, that glaciation during the Quaternary period resulted from lower temperature and more precipitation than prevailed in previous times. In my discussions of the several formations of the Tertiary period I pointed out that during Eocene time subtropical climate was widespread and that subsequently climate became progressively cooler and drier. Summer rains ceased and deserts appeared here in the West. Thus, the glaciation during the Quaternary period seems to have been the culmination of a long-term trend. The reasons for the changes are not understood although many intriguing explanations have been advanced, especially for the rapid fluctuations during the Quaternary period. It is clear, however, that the glaciation during the Quaternary period is not unique—extensive glaciations also occurred between 200 and 300 million years ago, at 450 and 700 million years ago, and at 3 billion years ago. One concludes from this that worldwide glaciations are rare events in earth history and must be the result of some special influences not normally effective between the widely spaced glacial episodes.

In the northern Sierra Nevada, glaciers occurred all along the crest as far north as the headwaters of Butt Creek. In the Diamond Mountains, small glaciers were present on the north slope of the north end of Grizzly Ridge overlooking Indian and Genesee valleys, on Keddie Peak, Kettle Rock, and on the north end of the Honey Lake fault scarp overlooking Janesville and Susanville. On the east side of Mt. Ingalls, a 3-mile-long glacier occupied the valley of Coldwater Creek and extended almost to Red Clover Creek. It was these valley glaciers, through their modification of the preexisting landscapes by erosion and deposition, that are responsible for much of the ruggedness and beauty of the region.

Valley glaciers, which are, of course, bodies of ice in motion, erode the rock over which they move and thereby modify an older landscape. When the ice melts and the glaciers disappear, the rock fragments contained in the ice are deposited, which also results in new features of the landscape.

Both the erosional and depositional results of glaciation are very well displayed in the Sierra Nevada between the Middle Fork of the Feather River and the North Fork of the Yuba River. That area, in Plumas and Sierra counties, most of which is now a recreational preserve, includes the well-known Plumas-Eureka State Park, the Lakes Basin, Gold Lake, and the spectacular Sierra Buttes. It is readily accessible by roads and trails and, excepting only Eureka Peak and McRae Meadow, is shown on the Sierra City quadrangle (15-minute series) of the U.S. Geological Survey. The discussion that follows centers on that area, and especially on that part which is tributary to Mohawk Valley—the drainage basins of Jamison, Gray Eagle, and Frazier creeks, shown on fig. 158.

To understand landscapes of glacial origin it is necessary to know how glaciers work, and that, in turn, requires some knowledge about water and ice. A most important property of water is that it expands on freezing—a given volume of water when frozen requires a space 9 percent larger. This increase in volume causes the ice to exert a great pressure on its container whether the container is closed or open. Freezing water breaks bottles, water pipes, automobile engine blocks and radiators, concrete roadways, and rock. People who live in cold countries understand this very well.

The joints or cracks in rocks are open containers, and when water in them freezes the sides are forced apart. The effect is very important even if the cracks are microscopic and the water only dew. In regions where glaciers occur, the disruptive effect on rocks is very significant, for even rock that appears to be hard and strong is fragmented in this way. This process, which is called frost wedging, is essentially continuous throughout that part of the year when thawing occurs during the day and freezing follows at night.

Another effect of this property is that ice will melt under pressure. One can experience this by clenching a piece of ice between the teeth—the teeth will be felt to penetrate the ice without breaking it. Or one can stretch a piece of wire across an ice cube and, by pressing downward, cut the ice in two without breaking it. This property permits some of the ice at the bottom of the glacier to melt, at least locally, because the pressure of the overlying ice changes as the glacier moves. A subsequent reduction of pressure will, for the same reason, cause the water to freeze again, so the disruptive effect on rock of alternate melting and freezing occurs beneath the ice as well as above it.

The ice of glaciers is not clean; it contains many angular pieces of rock that are acquired in several ways. The surface of the glacier becomes littered with pieces of rock that, having been loosened by frost wedging, fall from the rock walls that stand above the ice. Rock also moves onto the ice as rock falls and slides. Movement causes the ice to crack, and some of the rock at the surface falls into the cracks, in some instances as far as the bottom of the glacier. At the head of the glacier, motion of the ice in summer opens a great crack between the ice and the rock wall. Frost wedging is very intense there, and the loosened rock fragments fall to the bottom of the crack, where they become incorporated into the ice at the bottom of the glacier.

At the bottom and sides of the glacier, melting and freezing that result from pressure changes lift out of place blocks of rock that are surrounded by joints. This is a kind of quarrying operation called plucking. The plucked blocks, extracted, almost as though the glacier had roots or claws, are incorporated into the glacial ice.

A second important property of ice is that it is both ductile and brittle. It deforms and flows plastically even under the relatively small force of gravity. At the same time, it is brittle enough to crack, but its ductility permits cracks once opened to close and heal.

Because of its plasticity, or ductility, glacial ice moves under gravity from higher regions where it forms to lower and warmer regions. A glacier is analogous to a river—it is a stream of ice—and it is the motion of the ice that permits it to acquire its important load of contained rock, both by opening cracks and by plucking.

EROSION AND DEPOSITION BY GLACIERS

The beginning of glaciation is caused by a change of climate involving cooling and increased precipitation to such a degree that more snow falls during the winter than melts during the summer. Ice then forms by recrystallization of the snow. In a mountainous region, the first ice accumulates on a landscape that is the result of erosion of weathered rock by streams. The valleys of that landscape typically have a cross section in the shape of a V, and it is in them that ice forms and begins to move.

The moving ice quickly removes all vegetation, soil, and decayed rock and soon rests on fresh, unweathered hard rock. From the very beginning the ice contains some rock fragments, the supply of which will be maintained and augmented by the processes already described.

Rock fragments in the ice being dragged over the rock surface beneath and beside the ice abrade it and in turn are abraded. The valley is deepened and widened by the action of the glacier—more, perhaps, by frost wedging and plucking than by abrasion. The result is to change the cross-sectional shape of the valley from a V, typical of erosion by streams, to a U, typical of erosion by valley glaciers.

The best of the U-shaped valleys near the Plumas-Eureka State

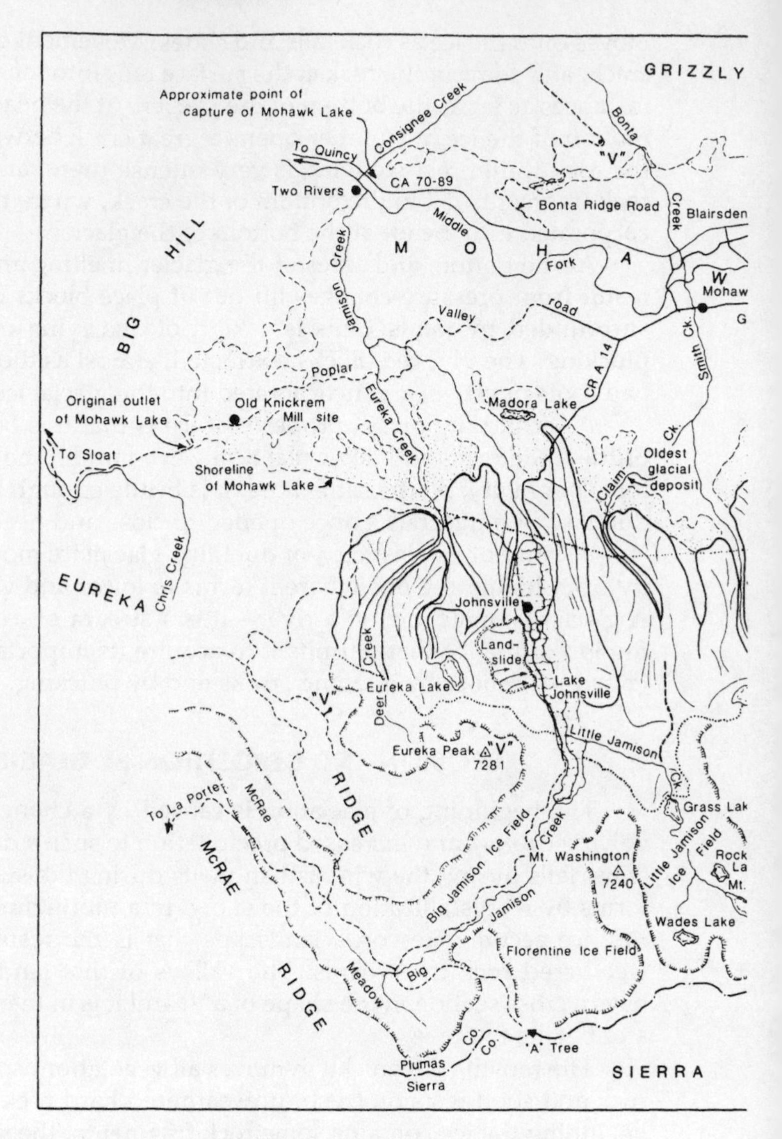

Fig. 158. Map of the glacial geology of the east front of the Sierra Nevada along the southwest side of Mohawk Valley.

Park are those of Little Jamison Creek between Mt. Elwell and Mt. Washington, shown in fig. 159; and of Jamison Creek and McRae Meadow, Gold Valley, and Lavezzola Creek. Others are the valleys of Frazier Creek near Frazier Falls and those occupied by Salmon, Packer, and Sardine lakes.

At their heads, glaciers excavate or quarry basins called cirques

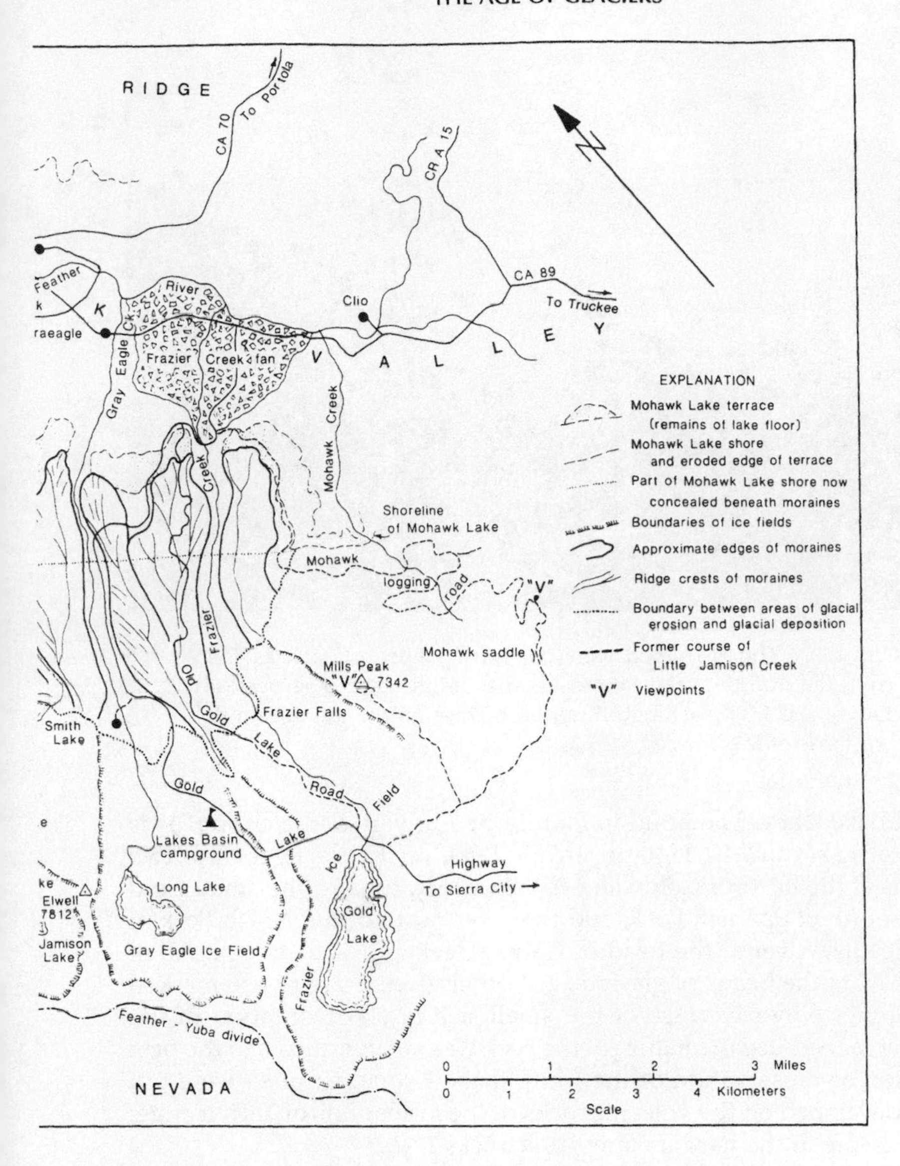

that are typically shaped like half of a bowl. Many cirques contain a lake after the glaciers have disappeared. In the most intensely glaciated regions, the cirque lake is entirely in solid rock (fig. 160), but in this region there is also much debris left from the melting of the last stage of glaciation. Distinctive cirques are those occupied by Wades, Rock, Jamison, Mud, Silver, Round, Young America, Upper Spencer,

Fig. 159. The U-shaped valley of Little Jamison Creek as seen from the trail between Johnsville and Smith Lake. The peak on the right is Mt. Washington, and the crags on the left are on the west side of Mt. Elwell.

and Snake lakes. The pond in Florentine Canyon is in a cirque, as is Crystal Lake on Mt. Hough; Taylor Lake on Kettle Rock; the two ponds at the head of Coldwater Creek on Mt. Ingalls; the small Gold Lake north of Spanish Peak; and the lakes north of the North Fork of the Feather River at the head of Chips Creek and Soda Creek. All of these mark the heads of glaciers, but other glaciers are not so marked by cirques either because of the small size or short duration of the glacier, or because the nature of the rock was not conducive to the production or preservation of the form of the cirque. Cirques were not well developed on the volcanic rocks of the formations of Tertiary age; the best are in the hard metamorphic rocks.

The back or head of the cirque is typically a cliff or nearly a cliff. Where two cirques nearly join, the intervening rock is often reduced to a nearly knife-edged ridge with a jagged skyline. The ridges that extend out to the northeast from Sierra Buttes are examples of such ridges.

The valley walls at the sides of glacial streams are steepened by glacial erosion, commonly to vertical, and in some instances they are left overhanging. After the glaciers have disappeared, such steep slopes are subject to collapse as massive rock falls, or to distintegra-

Fig. 160. Rock Lake from the west. It is a cirque lake that occupies a basin cut into and surrounded by hard rock. The downward slope to the left of the lake is the upper part of a step in the valley floor like that shown in fig. 162. Grass Lake lies below the step. Talus from the steep slope of Mt. Elwell is present both to the right and left of the lake.

tion by the fall of small pieces that build up conical heaps of loose rock called talus. A fine example of talus occurs on the southeast slope of Mt. Elwell beside Long Lake, as shown in fig. 161 (see also fig. 160).

Plucking beneath the glacier is most intense where joints or cracks are most closely spaced. For this reason different bodies of rock are subject to different rates of erosion. In places where a glacier passes in the downstream direction from a rock resistant to erosion because it has few joints to one much more easily eroded by plucking because it has more or more closely spaced joints, a steplike change in the slope of the rock surface beneath the ice may result. In some instances, because the riser of the step is vertical or nearly so, it becomes the site of a waterfall after the glacier disappears—this is the origin of Frazier Falls. Frazier Falls has a clear drop of about 200 feet, but the total height of the step, shown in fig. 162, is about 400 feet. Similar steps, although not so precipitous, occur below Rock Lake (see fig. 160), Long Lake, Upper Salmon Lake, and Upper Spencer Lake, and two steps are present in Big Jamison Canyon.

Glaciers erode not only by plucking but by grinding. The agents

Fig. 161. Long Lake and Mt. Elwell as seen from the Mt. Elwell trail west of the Lakes Basin Campground. The mass of loose blocks of rock that cover the lower slope of Mt. Elwell comprises a deposit known as talus. The blocks that make up the talus, some of giant size, have fallen from the cliff and slope above. At the left the lake water is against an almost vertical rock surface produced by the glacier that occupied the valley.

of grinding are the blocks of rock enclosed in the ice at the sides and bottom of the glacier. Moved over the rock surface under the weight of overlying ice, they grind away rock and leave smoothed, grooved, or scratched surfaces. On very hard rock the result may be a high polish. Of course, the blocks of rock that do the grinding are also affected in ways described below. Rounding and smoothing of the rock in the Lakes Basin, previously beneath the Gray Eagle glacier, show very well in fig. 163, a view looking southwest from the ridge east of the Lakes Basin Campground. The small rounded hills are called sheep-back rocks (from the French *roches moutonnées*). The downstream end of the sheepbacks are usually not smoothed but irregular, and show the effect of plucking. Figure 164 is a close view of a striated rock surface (see also fig. 57). The direction of scratches on the rock indicates the direction of ice flow. The upper limit of smoothing on the sides of a glaciated valley permits a measure of the maximum thickness of the glacier. For example, the upper limit of smoothing on the south side

Fig. 162. The barren rock across the center of the picture is the step in the valley of Frazier Creek over which the creek tumbles as Frazier Falls. This view is from the Old Gold Lake Road in the SE 1/4, Sec. 28, T. 22 N., R. 12 E. See also fig. 160.

of Eureka Peak at 900 feet above Big Jamison Creek shows that to have been the thickness of ice between Eureka Peak and Mt. Washington.

Because the prevailing temperature at any location depends on altitude, among other factors, the movement of glacial ice carries it downslope into regions warmer than that in which it formed. Indeed, some large glaciers extend into climatic zones far warmer than any in which ice could originate. Increased warmth means loss of ice and melting occurs not only at the top but along the sides and at the bottom of the glacier; some of the ice also evaporates (sublimates). Accordingly, glaciers are thinner and often narrower toward their lower end. The position of the lower end is determined as the place where forward advance is matched by melting. Because of climatic variations neither the rate of advance nor the rate of melting is constant, so glacial fronts advance and retreat from time to time. They do reach some maximum position of advance, and eventually they disappear completely.

Wherever the ice melts, the rock contained in it is deposited. Much or even all of the rock may be quickly removed by the stream of

Fig. 163. A view across the Lakes Basin from the ridge east of the Lakes Basin Campground that shows the rounded nature of surfaces eroded on hard rock beneath a glacier. This is the result of abrasion by the rock fragments carried in the ice. Mt. Elwell is on the right. The light-colored rock of the skyline ridge left of Mt. Elwell is that of the Sierra Buttes Formation. The Elwell Formation is not visible. All rock nearer the camera and of Mt. Elwell is of the Taylor Formation. The picture was taken from the same point as fig. 59, about 300 feet north of the S 1/4 Cor., Sec. 5, T. 21 N., R. 12 E.

meltwater that flows from the glacier, but in most instances a body of sedimentary rock recognizable as having been transported by a glacier remains. The varietal name of the aggregate of rock fragments is till when yet unconsolidated; after it becomes a hard rock it is called tillite. On occasion, however, blocks of rock may be left in odd places, such as perched on a peak or hill. Several such perched rocks, one of which is shown in fig. 165, are present in the Lakes Basin District.

Till is unsorted; huge blocks (fig. 166) and the finest rock powder, the result of grinding, are mixed together. The rock fragments, all of which are angular, are both those that were transported on top of the glacier and those frozen in the ice. Among the latter are those that were ground against the bedrock as the ice moved. They are easily recognized because they have surfaces that are grooved, or striated. Many are triangular in outline. Some striated surfaces are flat, but

Fig. 164. Glacial striae on rock of the Taylor Formation. By the trail to Mt. Elwell from the Lakes Basin Campground near its junction with the trail to Long Lake. See also fig. 57.

Fig. 165. Glacially perched rock south of Long Lake. A block of rock carried on or in the glacier and let down by chance on a peak when the ice melted.

Fig. 166. A very large glacially transported block of rock. Beside the Gold Lake Highway. The figure standing on it provides a scale. Sec. 29, T. 22 N., R. 12 E. See also fig. 58.

others are curved and resemble the sole of a well-worn shoe. Many striated blocks are chipped at their ends as if they had caught against rock in place and were fractured when they broke loose. These features of the blocks are shown in fig. 167. Blocks of this sort are abundant in road cuts in the till along the Old Gold Lake Road and the Gold Lake Highway, along County Road A-14 from Mohawk to Johnsville, between Johnsville and the ski area parking lot, along the road to Eureka Lake, and at Eureka Lake. Figure 168 shows the appearance of till on a large scale, and fig. 169 shows it close up.

Moraines are masses of till that have a distinctive form whose location bears a direct relationship to the glacier. Those ridges left along the sides of the glacier are called lateral moraines, and those that mark the maximum advance of the ice are terminal moraines. Complete melting of a glacier leaves a sheet of till called a ground moraine, which may be either pushed ahead of the next advancing glacier or overridden by it. All moraines are likely to be complex owing to the fact that the position of the end of the glacier is not stable, but alternately retreats and advances. Thus, the terminal moraine of a valley glacier tends to be a complex of arcuate ridges across the valley. During the final recession of a glacier, cold spells may reduce or stop the

Fig. 167. Blocks of rock abraded and broken in glacial transport. The flattened and striated surfaces, which are commonly triangular, and broken ends, such as the top of the smaller piece, are characteristic. Note the pencil for scale. Collected along the road from Mohawk to Johnsville. From the Tioga-age moraine on the east side of Jamison Creek.

Fig. 168. A cross section of a moraine. The till is unstratified and unsorted. Note the figure of a man at the lower center for scale. The pre-Sangamon moraine along the east side of Gray Eagle Creek exposed by construction of the Gold Lake Highway. NW 1/4, Sec. 28, T. 22 N., R. 12 E.

Fig. 169. A close view of the till of a moraine to emphasize its unsorted nature and the unrounded aspect of the larger pieces of rock. A road cut between Johnsville and the Johnsville ski area.

retreat for a while, at which times other arcuate moraines may be deposited across the valley; they are called recessional moraines.

Both terminal and the similar recessional moraines tend to be eroded away by the stream that flows across them after the ice has receded, but many remain as dams across the valley that retain the waters of lakes. Terminal moraines are not well preserved in this region, although most of the lakes are at least partially ringed by morainal deposits.

In contrast to the weakly developed and poorly preserved terminal moraines, the great lateral moraines left by the Big Jamison, Little Jamison, Gray Eagle, and Frazier glaciers are among the most conspicuous landscape features of the region. Eight moraines, one on each side of the four valleys, extend almost 3 miles into Mohawk Valley from the rocky heights of Eureka Peak, Mt. Washington, Mt. Elwell, and Mills Peak (see fig. 158). The largest is half a mile wide and 400 feet high. The upper, or southern, ends of each join the rock walls of the canyons through which the glaciers flowed. A considerable part of each was constructed by deposition on the floor of Mohawk Lake. Although the position of the preglacial lakeshore in the 6-mile stretch between Eureka Creek and Mohawk Creek is not precisely known because it is now covered by the glacial deposits, it is reasonably certain that from a half mile to a mile of each great moraine rests on or is partly embedded in the Mohawk Lake Beds.

These great moraines can be viewed from many places. They are in view from State Highway 70 near the top of the grade that leads down into Mohawk Valley. They can be seen to advantage from many high points on the Sierra Nevada, such as the Mills Peak lookout (see fig. 132), Eureka Peak, the ridge around Eureka Lake, or Eureka Ridge. The best view of all is that from a point on the Bonta Ridge Road, at the NE Cor., Sec. 4, T. 22 N., R. 12 E., altitude 5,560 feet, on the north side of Mohawk Valley. Unfortunately, the road is not good but the view is spectacular, as shown in fig. 150, which shows the moraine on the northwest side of Gray Eagle Creek resting on the Mohawk Lake Beds.

Some of the smaller glaciers of the region, such as those in the valleys of Deer Creek, Eureka Creek, and the eastern branch of Smith Creek, originated in a single cirque, but the larger ones were formed by the coalescence of ice streams that headed in several cirques. For example, the Big Jamison glacier was a composite of two ice streams, one from the poorly defined cirque at the south end of McRae Meadow, and one from the cirques in Florentine Canyon.

Each glacier, whether large or small, headed in an ice field. Figure 158 shows the ice fields from which ice moved toward Mohawk Valley. Minor ice fields and glaciers also occurred at the heads of Squirrel, Bear, and Mohawk creeks, and other unnamed creeks that head near Haskell Peak.

THE GLACIAL HISTORY

The four great glaciations recognized in eastern North America and Europe were responses to worldwide changes in climate. Therefore, glaciations may be expected to have occurred at about the same time in all the colder parts of the earth, and, to be sure, four major advances and retreats of glaciers are now known to have occurred along the east side of the Sierra Nevada in the region between Bridgeport and Bishop. They are recognized by the deposits of till that remain.

Since the four were first recognized in California, several other minor glaciations have been proposed for the Sierra Nevada, some of which remain controversial. The ages of the Sierran glaciations do not correspond exactly to those recognized in eastern North America, but some discrepancies should be expected because of the great width of the continent and the intervention of mountains, which have a pronounced effect on climate. Of course there may be other reasons in a matter as complex as large variations in climate. The glacial chronology is summarized in fig. 170.

In the region between Bridgeport and Bishop, the till of the two oldest glaciations—the McGee, older than 2.6 million years, and the Sherwin, younger than McGee but older than 700,000 years—lacks form; that is, because of the extended period of weathering and erosion since their deposition, no recognizable moraine ridges remain. The deposits of the next younger glaciations—the Casa Diablo and the Mono Basin—lack distinct ridge crests and the rocks are strongly weathered. The Tahoe and related Tenaya stages, which had their maximum extents of ice in the interval of 45,000 to 75,000 years before the present, are obvious moraines, but the rocks of the till are also somewhat weathered, the crests of the moraine ridges are rounded, and terminal moraines are obscure or absent. In contrast, the deposits of the next younger—the Tioga glaciation—are sharp-crested lateral moraines with well-defined terminal or recessional moraines, usually inside of and smaller than the moraines of the Tahoe and Tenaya glaciations. The only slightly weathered Tioga deposits are the result of a maximum extent of ice about 20,000 years before the present.

Three younger ice advances of small extent are now also recognized. The Hilgard is dated as 9,000 to 10,500 years before the present. The Recess Peak is dated at 2,000 to 2,600 years before the present, and the Matthes at 700 years before the present, which was during the Middle Ages. These names of till deposits are used in the sense of formation names.

A study by Birkeland in the Lake Tahoe–Truckee region resulted in recognition of a till without form that he named the Hobart, and that he thought to be no more than a million years old, and moraines equated to the Tahoe, Tioga, and Hilgard glaciations.

The ages assigned to the glaciations are taken to be the same as the ages assigned to the associated deposits of till, which may be determined in many ways: potassium-argon dating of igneous rocks or volcanic ash below, within, or above the till; radiocarbon dating of wood or charcoal found in the till; the state of preservation of the form of the moraine; the degree of weathering of certain kinds of rock in the till; and even the size of lichens growing on the exposed blocks of rock of the till.

The glacial deposits of Plumas and Sierra counties, not previously examined in detail, are the subject of a study not yet published by S. A. Mathieson. Mathieson found no igneous rocks, volcanic ash, or wood in the till that could be used for absolute dating, but he has found other means of establishing ages, as explained further on.

GLACIAL CHRONOLOGY

	EASTERN NORTH AMERICA			CENTRAL SIERRA NEVADA		PLUMAS AND SIERRA COUNTIES	
Epoch	Glaciation	Interglacial	Years before present (Approx.)	Glaciation	Years before present (Approx.)	Glaciation	Years before present (Approx.)
			0				
Holocene				Matthes Recess Peak Hilgard	700 2.000-3.000 9.000-10.500		
			10.000				
Pleistocene	Wisconsin			Tioga Tenaya Tahoe	20.000 45.000 (?) 60.000-70.000	(Tioga) II Post- Sangamon I	8.000--20.000 60.000-70.000
			15.000				
		Sangamon					
			130.000				
	Illinoisan			Mono Basin	130.000 +	(Tahoe) IV III Pre- II Sangamon I	140.000 150.000 170.000 190.000
			270.000				
		Yarmouth					
			350.000				
	Kansan			Casa Diablo	400.000		
			600.000				
		Afton					
			1.000.000				
	Nebraskan			Sherwin McGee Deadman Pass	700.000 2.600.000 2.700.000		
			2 to 3 million				
Pliocene							

Fig. 170. A summary of the glacial chronology of the Sierra Nevada.

That which I believe to be the oldest till of the region is visible along Claim Creek, which is the northwesterly branch of Smith Creek, in the SW 1/4, Sec. 17, T. 22 N., R. 12 E., a mile and a half southwest of Mohawk. There, buried in Mohawk Lake sediments at an altitude of 4,600 feet, is typical till that consists of blocks of rock as long as 4 feet with flattened and striated faces set in an unsorted matrix of smaller angular rock fragments down to the size of powder. This occurrence is about 400 feet lower than the shoreline of Mohawk Lake, and about 200 feet lower than the floor of the lake when it was drained. It is about a mile into the lake from the now concealed preglacial shoreline (see fig. 158). Although I cannot even estimate the age of this till, I believe it to be considerably older than the next younger till, which appears to rest on the Mohawk Lake terrace.

The next younger till is that which is a part of the great lateral moraines along Jamison, Claim (the west branch of Smith Creek), Gray Eagle, and Frazier creeks that have already been described in general terms, and of the moraines farther south along Sardine, Packer, Sawmill, and Salmon creeks. All of these are, however, more complex than has so far been indicated in that each consists of an older and a younger part that are easily distinguished by the color of the till.

Each of the great moraine ridges consists of a set of approximately parallel smaller ridges. The till of the older ridges is weathered brown, the constituent rock fragments are decayed to such an extent that their characteristic shape is often difficult to recognize, and striated surfaces are absent or difficult to find. The originally sharp ridge crests have been rounded and subdued by weathering and erosion.

The younger tills comprise ridges that are still sharp crested, and their color is gray to greenish gray. Flattened and striated surfaces on the constituent fragments are abundant, well preserved, and easily found. The moraine ridges of the younger till are "inside" those of the older till; that is, they are nearest the stream that drains the valley between the great moraine ridges.

These differences between the older and younger till and ridges are readily seen along the road from Mohawk to Johnsville and along the Gold Lake Highway.

At the top of the mile-long steep grade between Mohawk and Johnsville, a nearly flat stretch of ground that is a part of the floor of Mohawk Lake is marked by the sign that designates the boundary of the Plumas-Eureka State Park. The road crosses the flat and then climbs over a low ridge that is the outermost ridge of the great moraine on the southeast side of Jamison Creek. Just beyond the ridge is the parking area at the trail to Madora Lake.

The ridge consists of brown older till, the blocks of which lack striations because they have been destroyed by weathering. The ridge continues to the north toward Madora Lake but it does not connect with any recognizable terminal moraine.

Farther toward Johnsville, not more than 0.2 mile beyond the parking area, the till beside the road is not brown but is the gray to grayish green of the younger till—much less weathered than the older one. Striated blocks are easily found. The younger till comprises a series of sharp-crested moraine ridges inside the older moraines. That these younger ridges curve to the northwest, toward Jamison Creek, can be seen on the west side of the road in the W 1/2, Sec. 18. Partly obscured by trees along the road and supporting a luxurious growth of brush, they are not very conspicuous. The curved ridges are terminal moraines that mark a series of halts during the recession of the glacier. Their central parts have been eroded away by Jamison Creek, but their continuations are present on the opposite side of the valley. They once dammed the valley with results described farther on.

The distinction between older and younger till and moraines is also well displayed on the Gold Lake Highway in the NW 1/4, Sec. 28, T. 22 N., R. 12 E. The place is best found by measuring distance from the bridge across Frazier Creek. For 0.3 mile from the bridge to the junction with the Old Gold Lake Road, the cuts are in Mohawk Lake Beds. For the next 1.7 miles the cuts are in brown, weathered rock that is a mixture of old till and soil. The bench on the north side of the road is part of the floor of Mohawk Lake. The significant location is where the road comes out of forest into brushland. On the left (south) is a ravine, and on the right is a small knob with a few trees standing in the brush. On the east side of the ravine is the tall cut in brown weathered older till, shown in fig. 168. On the west side of the ravine, and around the turn of the road beyond the ravine, is gray to grayish green till of the younger moraine, also of the Gray Eagle glacier. Directly ahead is the sharp-crested younger moraine on the opposite (northwest) side of Gray Eagle Creek. From here toward Gold Lake the road is on younger till as far as the entrance road to Gray Eagle Lodge. In that interval the road crosses about a dozen sharp-crested recessional moraine ridges that both point and slope toward the bottom of the valley. At the entrance to Gray Eagle Lodge there is a tall road cut in the younger till that is an excellent place to examine its character.

It might be thought from these descriptions of the older and younger moraines that they correspond to the Tahoe and Tioga glaciations in the region between Bishop and Bridgeport. However, Mathieson has shown that the glacial history is still more complex. As is the case elsewhere in the world, each of the glaciations thus far mentioned involved lesser advances and recessions in response to changes of climate. By careful and detailed observations, Mathieson has shown that the older glaciation is a complex of four advances and recessions and that the younger glaciation involves two advances and two recessions. He has also been able to assign tentative ages to these several stages of glaciation by means briefly explained in the next paragraph.

It has been known for several decades that the ratio of the isotopes of the element oxygen, ^{18}O and ^{16}O, secreted into the calcareous shells of marine organisms depends on the temperature of the water in which the organisms lived. Hence the water temperature at the time that the shell was formed can be determined by measuring the amounts of the isotopes in shells.

Studies of marine organisms that lived during the Quaternary period have permitted the recognition of sequential temperature changes in sea water that are believed to correspond to the advances and recessions of glaciers on land and to the several interglacial episodes. The ages of some of these changes have also been established by means of radiometric age-dating techniques. In this way an absolute time scale of temperature changes has been established.

By comparing the relative ages and the relative durations of the several glacial episodes in Plumas and Sierra counties with the marine temperature-time scale, Mathieson was able to establish probable ages of the local glacial advances.

Thus he has found that the four advances of the older glaciation occurred at about 190,000, 170,000, 150,000, and 140,000 years before the present. All fall within the Illinoisan, the next to last of the four great glaciations in North America. The interval between the early and later glaciations lasted about 60,000 years and corresponds to the Sangamon interglacial, the last long interglacial episode.

The two advances of the younger glaciation occurred during the last 80,000 years, and both are in the Wisconsin, the last great glaciation. The first is thought by Mathieson to have occurred in the interval of 80,000 to 62,000 years before the present, and the youngest between 26,000 and 8,000 years before the present.

Mathieson recognizes the difficulties in correlating the several glacial advances in Plumas and Sierra counties with those farther south, and he prefers to designate the former simply as pre- and post-Sangamon. However, the last advance, between twenty-six thousand and eight thousand years before the present, seems to both Mathieson and to me to correspond quite well with the Tioga glaciation of the more southern region. I prefer to use that name for the last glaciation here in spite of the fact that the youngest deposits cannot be easily

Fig. 171. Evidence of multiple glaciations. A younger post-Sangamon moraine, the higher one, crosses a pre-Sangamon moraine. The lower, older moraine is weathered and has a nearly complete cover of brush. Much relatively unweathered rock shows through the open vegetation on the younger moraine. These are the moraines on the south side of Sardine Creek and Lower Sardine Lake viewed from the Gold Lake Highway between Bassetts and its junction with the road to Sardine Lakes.

distinguished from the next older. The glacial history set forth above is summarized in fig. 170.

Illustrative of the difficulties in correlating glacial deposits is the fact that Mathieson believes that the last of his pre-Sangamon advances, about 140,000 years ago, is equivalent to the Tahoe advance of the central Sierra Nevada, which other people consider to be post-Sangamon and less than 90,000 years old. Obviously the situation is muddled and I have no personal contribution to make toward resolving the discrepancy. Figure 171 should set to rest any doubts concerning the reality of multiple glaciations. It shows a post-Sangamon moraine that appears to rest on top of a pre-Sangamon moraine.

I pointed out earlier that the younger moraine ridges crossed by the road from Mohawk to Johnsville curve toward Jamison Creek and that corresponding ridges are present on the opposite bank. These ridges constituted a terminal moraine dam across Jamison Creek. A body of water that I call Lake Johnsville accumulated behind the dam.

Fig. 172. The very flat surface of the sedimentary fill of Lake Johnsville. At the cemetery in Johnsville.

The lake was about 2.5 miles long but only about a quarter mile wide. Its head was about a quarter mile south of the state park campground. The basin became completely filled with sediment, the dam was breached by Jamison Creek, and a good deal of the sediment has been removed. However, much of the flat surface of the fill remains, and Johnsville is situated on that surface (fig. 172). The nature of the sediment—coarse grained, poorly sorted, poorly bedded, and gray to gray green in color—is visible in the banks of Jamison Creek, both above and below the bridge at Johnsville (fig. 173). Obviously the lake was formed in connection with the later glaciation, and probably with the latest, or Tioga stage. The surface elevation of Lake Johnsville was close to 5,200 feet, about 160 feet higher than that of Mohawk Lake.

Further complications in the Jamison Creek basin involve the relationships of the Little Jamison and Big Jamison glaciers. As I have indicated, the Big Jamison glaciers, fed by a large ice field, excavated the valley and left moraines of both ages. Although Little Jamison Creek is now tributary to Big Jamison Creek, the glaciers in those two valleys did not join. Before the post-Sangamon glaciation, the headward valley of Little Jamison Creek was continuous with that of Claim Creek, which is the west branch of Smith Creek. That this was so is indicated by the presence in that almost inaccessible valley (in Secs. 18 and 19, T. 22 N., R. 12 E.) of moraines that, according to Mathieson, belong to both the pre- and early post-Sangamon advances. In the light of this

Fig. 173. The sedimentary fill of Lake Johnsville exposed in the bank of Jamison Creek at the east end of the bridge at Johnsville. This youthful sediment has the same grayish green color characteristic of the post-Sangamon moraine that partly surrounds the basin.

fact, the great moraine ridge between the valleys of Big Jamison Creek and Claim Creek must be a composite of the deposits of the several glaciers exclusive of the Tioga. The diversion of the headward part of the Little Jamison Creek below Grass Lake (fig. 174) may have occurred either before or after the youngest (Tioga) advance, but the manner by which the diversion was accomplished remains uncertain. The moraine ridge between the two drainage basins left by the pre-Tioga glaciations no doubt once extended to Mt. Washington. However, moraine ridges are ordinarily not of uniform height, and there may have been a low region, a saddle, on the ridge where the diversion occurred. A recessional moraine across Little Jamison Creek or a terminal moraine in front of advancing Tioga ice might have dammed the valley to create a small lake, which may then have found its outlet, not down Claim Creek as before, but across the saddle on the moraine ridge into Lake Johnsville. Erosion by the diverted stream down the steep slope would have rapidly cut the present gap through the moraine and removed all evidence of the existence of the lake.

An alternate possibility, and perhaps the most likely one, is that a section of the moraine ridge simply slid into Lake Johnsville and the stream followed through the gap. Landslides do occur in till because of its unconsolidated nature. In fact, a slide almost a quarter of a

Fig. 174. The course of Little Jamison Creek diverted toward the camera into Big Jamison Creek, viewed from the slope of Eureka Peak across the valley. The remaining Jamison Mine buildings are in the lower left corner of the picture.

square mile in area is present on the slope on the opposite side of Jamison Creek below Eureka Lake (see fig. 158). Its toe rests on the flat surface of the sediment that filled Lake Johnsville between the Plumas-Eureka Park headquarters and Johnsville. Other slides of till are present along Frazier Creek.

Both pre- and post-Sangamon moraines are present along Frazier Creek. The till along the Old Gold Lake Road, in Sec. 32, T. 22 N., R. 12 E., is weathered brown and no doubt is pre-Sangamon. Farther north, in Sec. 33, the till along the road is gray to gray green and is evidently post-Sangamon (Tioga). The crest of the ridge is sharp, and striated rock fragments are abundant. The ridge on the east side of Frazier Creek is very complex, and Mathieson believes that the older of the post-Sangamon glaciers was diverted eastward into the basin of Mohawk Creek.

Probably it was the pre-Sangamon glaciers that were primarily responsible for the deep excavation of the valley that left the great step at Frazier Falls (see fig. 162). That the pre-Sangamon glaciers extended into Mohawk Lake is evident because of the presence of brown-colored weathered till along both the Gold Lake Highway on the northwest side of the creek and the Mohawk–Chapman logging road on the southeast side. On both sides it appears to rest on the Mohawk Lake terrace at about 4,800 feet, well below the water level of the lake.

The Tioga-age glacier in Frazier Creek extended almost as far because its till is present below the 5,040-foot level of Mohawk Lake only a quarter mile upstream from the crossing of Frazier Creek by the Gold Lake Highway.

The Frazier glaciers were derived mostly from the Gold Lake basin. There was also a large ice field east of Gold Lake that not only contributed to the Frazier glaciers, but sent ice southward down both Howard and Salmon creeks. Disappearance of this ice resulted in a sheet of till that covers between 2 and 3 square miles east of Gold Lake. Hollows on the irregular surface of the till are occupied by Snag, Haven, and Goose lakes and by other ponds nearby.

The presence of moraines of two ages is not confined to the four main basins. Both are also present along Deer Creek, the first large basin west of Johnsville, and along Eureka Creek. A small glacier of pre-Sangamon age that originated on the north side of Eureka Peak left two moraine ridges and a sheet of weathered till over the slope below Eureka Lake where the Johnsville ski tows are located. The moraine ridge that closes the Eureka Lake basin (fig. 175) is apparently of Tioga age.

In the same way, pre-Sangamon glaciers that formed on the north slope of Mt. Elwell spread lateral moraine ridges and a sheet of till over the lower slopes. A younger Tioga-age glacier was at least in part responsible for the moraine that contains Smith Lake.

Glaciers of both ages occurred also in the basins now occupied by Salmon, Deer, Packer, and Sardine lakes, and also in the valleys tributary to the North Fork of the Yuba River east of Sierra City.

Three tills younger than Tioga have been recognized in the central Sierra Nevada, and one has been recognized north of Truckee. Their names and ages are listed in fig. 170. I am unable to identify with certainty any younger tills in the Plumas-Sierra County region. Although some lesser bodies of till are present, they could have resulted from the last stages of the Tioga glaciation.

A small lateral moraine is present along Big Jamison Creek just upstream from the mouth of Florentine Canyon. Small terminal or recessional moraines are partly responsible for retaining the waters of Silver and Round lakes and the ponds called Mud and Helgramite lakes on the west side of Long Lake. These might have been the result of small glaciers younger than the Tioga, but I have no evidence that they are.

Till is present also near the outlets of both Long and Gold lakes.

Fig. 175. Eureka Lake as seen from Eureka Peak. The sparsely vegetated ridge at the far end of the lake is a terminal moraine of post-Sangamon or Tioga age. The densely vegetated moraine at the right is the pre-Sangamon moraine of the west side of Jamison Creek.

Because till in such positions requires that both lake basins be filled with ice, it probably represents a stage in the disappearance of the Tioga-age glaciers.

In summary, six and possibly seven glacial advances are recognizable among the till deposits of the crestal region of the Sierra Nevada in Plumas and Sierra counties between the Middle Fork of the Feather River and the North Fork of the Yuba River. The two main glaciations—Mathieson's pre- and post-Sangamon—are responsible for modification of the landscape by both erosion and deposition to the extent that major landscape features resulted. Of course, they are at the same time responsible for much of the scenic beauty of the region.

I have already indicated that glaciation also occurred elsewhere in the northern Sierra Nevada and on the Diamond Mountains, but I cannot even suggest what the history might be in those places. Much of what I have written here about glaciation is still tentative and surely incomplete.

There remains, however, the problem of the relationship of the glacial deposits to those of Mohawk Lake, because, to the degree that the ages of the glacial deposits are valid, they give some indication of the time of disappearance of Mohawk Lake.

To address that problem we turn again first to the relationships along County Road A-14 between Mohawk and Johnsville, and in the vicinity of Madora Lake.

The altitude of the Mohawk Lake floor, or terrace, at the sign that marks the boundary of the Plumas-Eureka State Park is 4,890 feet, or 150 feet lower than the probable 5,040-foot level of Mohawk Lake. Gravel of the lake floor is present here and also on the other side of the ridge of pre-Sangamon moraine. Furthermore, one can walk through the woods toward Madora Lake and pass around the end of the moraine ridge, all the while remaining on the gravel of the lake floor. Because of this it appears that the pre-Sangamon till was deposited on the lake floor below the level of the lake. In other words, the lake was present when the pre-Sangamon glaciation occurred. One would feel more certain about this if a shoreline or beach could be identified on the moraine, but that has not been accomplished, principally because of the presence of dense brush.

However, there are indications of the shoreline on the pre-Sangamon moraine on the west side of Gray Eagle Creek, a mile west of Graeagle (in the NE 1/4, Sec. 21, T. 22 N., R. 12 E.). Here the pre-Sangamon till, identified by its brown color and the smoothly rounded ridges, rests on the Mohawk Lake terrace. In 1950, a few years after a wildfire had cleared the brush and trees, I made a careful survey of the area. Well-rounded gravel occurs on the brown-colored pre-Sangamon till as high as 5,040 feet, and a fairly well defined gravel-covered bench occurs at an altitude of 4,866 feet. Since this was a rocky point extending into the lake, one would not expect to find a well-defined beach, but the evidence seems clear that the moraine was constructed on the lake floor in water about 175 feet deep. Accordingly, it is possible to say with assurance that Mohawk Lake existed until sometime during the period of 130,000 to 75,000 years ago, the estimated age span of the Sangamon interglacial.

From the known relationship of Tioga-age (post-Sangamon) moraines to the lakeshore it is not possible to be certain that the lake still existed at that time. Along Jamison Creek, the Tioga-age glacier built the terminal moraines that retained the waters of Lake Johnsville, whose surface altitude of 5,200 feet was 160 feet higher than that of Mohawk Lake. Because of the difference in altitude of the two lakes, the presence of the terminal moraines, and the extent of later erosion, it is not now known that the Tioga-age glacier extended below the 5,040-foot level. Thus there is no evidence along Jamison Creek that Mohawk Lake still existed.

Evidence is also lacking along Gray Eagle Creek, but it is probable that the Tioga-age glacier terminated short of and above the Mohawk Lake terrace.

Along Frazier Creek, young till is present below the 5,040-foot level of Mohawk Lake; therefore, it is possible that the Frazier Creek glacier of Tioga age entered the lake perhaps only 20,000 years ago. The subject is pursued further below.

THE TIME OF DRAINAGE OF MOHAWK LAKE

The last ten thousand years of geologic time was previously called the Recent; it is now the Holocene epoch of the Quaternary period. The Hilgard and the younger glacial advances occurred during the Holocene, but neither Mathieson nor I have been able to identify for certain any till of that age. If the drainage of Mohawk Lake occurred during or after the Tioga glaciation, as I believe, it was an event of the Holocene.

Mathieson does not agree with my view. His opinion is that Mohawk Lake was drained before or during the Sangamon interglacial, not less than 75,000 years ago, and that the entire post-Sangamon glaciation came after the drainage. I feel that Mathieson has not proven his concept, and yet I have no positive proof to the contrary. There is, however, an approach to the problem that points to the more recent date.

During the forty years ending in 1978, I spent most summers in Mohawk Valley and crossed the Middle Fork of the Feather River by way of the Mohawk Bridge countless times. Most times I noted the condition of the river, and although I made no surveys I estimate that the bed of the river at the bridge was lowered by 2 to 3 feet during my period of observation. Possibly most of that occurred during the very high water in 1963. I am quite certain that the amount of lowering is more than 2 feet, but to present a conservative view I assume the smaller value. During forty years that represents an average annual lowering of 0.05 foot.

Taking the altitude of the Mohawk Lake floor before the drainage of the lake to be 4,866 feet, its height a mile west of Graeagle, which is quite near the center of the lake, and the altitude of the present river near Blairsden as 4,320 feet, the amount by which the river has lowered its bed by erosion since the drainage amounts to 546 feet. At the average annual lowering rate computed above, this could have been accomplished in 11,000 years, half the time since the Tioga glaciation.

It is instructive to compare this estimate with Mathieson's concept of drainage of the lake before the end of the Sangamon interglacial, which was about 75,000 years ago. The lowering of the valley floor of 546 feet during that time interval yields an annual rate of 0.007 foot, or less than three thirty-seconds of an inch. At this rate, the lowering of the river bed during my forty years of observation would have amounted to less than 4 inches, an amount so small that I could not have observed any change.

There are many reasons why erosion might not have been continuous or its rate constant, and, after all, I have only an eye estimate; nevertheless I consider my observation to be a very strong indication that the draining of the lake occurred during or after the Tioga glaciation was completed and that it may well have been more recent than 20,000 years ago. I believe that it did not occur as long ago as the Sangamon interglacial.

SMITH CREEK TERRACE

Underlying the Mohawk Cemetery and exposed in the cut bank of County Road A-14 and also in the cliff that borders the river is a body of coarse gravel of well-rounded cobbles that rests unconformably on fine-grained Mohawk Lake Beds. This and other patches of similar gravel that occur to the southwest in the valley of Smith Creek are remnants of a terrace deposit left by an earlier Smith Creek that flowed on the valley floor when the floor was about 80 feet higher than it is today. For the gravel to accumulate there must have been a cessation for a short time in the deepening of the valley. The obvious cause of such a halt is the damming of the river by either one or both of two landslides, each about a half square mile in area, on the north side of the Middle Fork in the canyon between Mohawk Valley and Sloat. Both are marked along State Highway 70-89 by the presence in the road cuts of abundant white cobbles and a number of very large unrounded masses of green metamorphic rock (see fig. 156).

The origin of the steep-walled canyon, the channel through which Mohawk Lake was drained, has already been described. The steeply sloping walls of the canyon, cut in the soft rock of the Penman Formation in an area where there are many faults, were left only poorly supported. The normal tendency of soft rock on a steep slope to slide was enhanced by the presence in the canyon wall of an old river channel filled with gravel, partly of the Auriferous Gravel and partly of gravel at the base of the Bonta Formation. The gravel is unconsolidated, it has a matrix of sandy clay, and, being cut off as it was by a fault parallel to the river, it was saturated with water. All are factors ideal for the development of a landslide. In fact, nearly everywhere where gravel is present in the mountain slopes landslides have occurred. That the slide farthest upstream did dam the river is proven by the presence of some of the slide material still remaining on the south side of the river.

Such a dam no doubt produced a small lake, probably no more than 150 feet deep and about 3 miles long, in the lower part of the Mohawk Valley. Mud, sand, and gravel would have accumulated in the lake, and coarser gravel would have accumulated in the stream beds upstream from the lake. Once the small basin was filled and water spilled over the dam, it and the accumulated sediment were quickly removed by erosion. It is the remnants of the coarser gravel deposited along the course of Smith Creek that constitute the terrace deposits at the Mohawk Cemetery and elsewhere nearby. This entire event probably took place within the last few thousand years.

FRAZIER CREEK FAN

The last significant geologic event in the Mohawk Valley was the deposition of the rocky pile upon which most of the beautiful and pleasant subdivision known as Graeagle is situated. I call the deposit the Frazier Creek fan because it is triangular or fan shaped and its origin is related to Frazier Creek. The stem or head of the fan, the highest point on it, is at the mouth of the canyon of Frazier Creek at the bridge where it is crossed by the Gold Lake Highway. The northwesterly tip is at the point where Gray Eagle Creek enters the Middle Fork, and the other tip is on the Middle Fork where State Highway 89, proceeding toward Clio, leaves the forest and drops 20 or 25 feet to the level of the grassy meadow of Mohawk Creek. The stony surface of the fan once extended to the Middle Fork, but most of that part north of State Highway 89 was reclaimed for the golf course and residences.

The deposit clearly rests on Mohawk Lake Beds. It surrounds a small hill of Mohawk Lake Beds just west of Gray Eagle Creek on the south side of State Highway 89, and it buried two small hills of Mohawk Lake Beds that are now exposed beneath it in road cuts along the Gold Lake Highway between its junction with State Highway 89 and the Frazier Creek Bridge.

The thickness of the fan material over those hills is less than 10 feet. Along its south edge, borings in connection with a once-proposed dam show its thickness to be not more than 25 feet; and elsewhere pits for septic tanks have penetrated through the fan. It is very thin. Obviously the fan was deposited on an erosion surface like that presently on both sides of it, and at the same altitude.

Unlike the terrace gravels at Mohawk, the rocks of the fan are not rounded. They are distinctly angular and the deposit is unsorted. All sizes are jumbled together; pieces as much as 6 feet in largest dimension are embedded in successively finer material like that of till (fig. 176). A few moments' search will, in fact, reveal blocks that have flattened and striated surfaces and broken ends that are the characteristic marks of transportation by glaciers. There is little doubt that the material of the fan was once till, and its quite unweathered character indicates that much if not all of it was of Tioga age.

Fig. 176. The material of the Frazier Creek fan disturbed by road construction. It consists of unweathered angular blocks of rock, many of which have flattened striated faces and broken ends like those shown in fig. 167. The color is grayish green like the till of the Tioga-age moraines from which it evidently was derived. In Graeagle.

Of course, the fan was not deposited by a glacier. No glacier existed so recently or at such a low elevation. The facts that the surface of the fan slopes radially downward from its head at the mouth of the Frazier Creek Canyon, that the largest blocks at any place on the fan are smaller in proportion to their distance from the head of the fan, and that Frazier Creek flows downslope across it indicate that the material came down Frazier Creek. It is the kind of material that could have resulted from the bursting of a dam across Frazier Creek and that is probably what happened, although the full story is not yet known.

What is known is that there are fairly extensive landslides of till from the moraines on both sides of Frazier Creek, especially in Sec. 33, T. 22 N., R. 12 E. These can be seen from the Old Gold Lake Road. What appears to have happened is that one or more of these slides, initiated perhaps at a time of torrential rain, dammed the creek for a short time, and when sufficient water had accumulated behind the dam it burst, as have some manmade dams. The resulting flood of rock, mud, and water, truly a mudflow, rushed down the canyon and spread over the floor of Mohawk Valley. It spread to the east across part of the meadow of Mohawk Creek and created another dam across the Middle Fork of the Feather River a half mile downstream from Clio (fig. 177). The entire process could have been completed in a few

Fig. 177. The upper or southeastern edge of the Frazier Creek fan as seen looking northwesterly from State Highway 89 a half mile west of Clio. The line between grassland and forest is the boundary between the fan deposit and Mohawk Lake Beds. The site of the damming of the river by the fan is at the right side of the picture where the river now flows in a narrow channel confined between the fan and hard rock on the right bank. The foreground is a meadow along Mohawk Creek that is known from borings made in exploration for a dam proposed for this site to be underlain by fine-grained Mohawk Lake Beds.

hours' time. An alternate possibility is that the dam across Frazier Creek was a terminal moraine left by the Tioga-age glacier. The bursting of a moraine dam would have the same effect as one that had resulted from a landslide.

The transportation and deposition of the mudflow composed of till happened so quickly that the characteristic features of prior transport by glacial ice were unaffected.

The effect on the Middle Fork of the Feather River was extensive. When the river overtopped the new dam, its new course from below Clio to its confluence with Gray Eagle Creek was constrained to skirt the edge of the mudflow along the northeast wall of the valley, where it remains today. The dam created a lake at least 30 feet deep that extended up Sulphur Creek no less than 2 miles. The lake then filled with sediment, mostly gravel. The Middle Fork has gradually cut through the dam, and most of the fill of the lake has been removed. That which remains forms the distinct terrace near Clio, shown in fig. 178.

Fig. 178. Mohawk Valley looking southeasterly from State Highway 89 at a point only a few tens of feet from that at which fig. 177 was taken. This is the site of the lake that formed behind the dam across the Middle Fork of the Feather River at the upstream edge of the Frazier Creek fan. The terrace that resulted from the filling of the lake basin with sediment and the subsequent partial removal of that sediment by erosion is the flat surface visible through the trees to the right and its continuation into the far distance. The terrace surface is underlain by the gravel that filled the basin. The meadow to the left of the terrace is on a younger erosion surface cut into the lake deposits by the Middle Fork and Sulphur Creek since the dam was breached.

An approximate age of the fan can be determined by applying the rate of lowering of the valley floor used above. The erosion surface beneath the fan slopes toward the Middle Fork. The altitude of the lowest point on that surface, which is near the junction of Gray Eagle Creek and the Middle Fork, is about 4,350 feet, which is 30 feet higher than is the river at Blairsden. At the rate of lowering of the valley of 0.05 foot per year, the erosion surface might have been established at that level as recently as 600 years ago. This may be taken to be the age of the fan, for erosion of the surface was terminated by the deposition of the fan upon it. Such a computation may be badly in error for such a short period of time but in any case the deposition of the fan was a very recent event. It seems to have been the last significant geologic event to have occurred in this region.

Epilogue

And so in 1985 I have finished this account of the events of the last 400 million years as revealed in the northern Sierra Nevada. It is still an incomplete account for much remains to be done. Yet it is more complete in many respects than it could have been only twenty years ago because much of our present understanding depends on the concept of plate tectonics, a true revolution in geological thought that began as recently as the 1960s. Although the underlying causes of plate motion are not yet completely understood we do seem to know, for example, the proximal causes of the origin of folded mountain systems.

In this history I have related how the deep ocean floor became a site of deposition of sediment and volcanic material, how those deposits became a part of the continent, how plutons formed and entered into the mountain structure, and how the mountains were elevated and then eroded away. This chain of events was not unique to the Sierra Nevada; it is one that has happened repeatedly in earth history. It belies the common literary reference to the "eternal mountains": a better aphorism would be "nothing is eternal—change is constant."

The earth is a heat engine driven by both heat from its interior and that received externally from the sun. The heat reserve, both internal and external, is very large, and so change will continue into a future whose end may be foreseen only by cosmologists.

The rates of change of the earth are small in terms of the expected human life-span. Yet changes in elevation or depression of the earth surface are measurable, and, as noted in the case of Lake Mead, some are even caused by man. Changes owing to erosion and deposition are measurable, and some of those are manmade. The movement on some faults is being measured with the hope that such studies may lead to the prediction of earthquakes.

The development and use of instruments for measuring such changes, for chemical analysis, and for understanding the interior of the earth are contributing to a better comprehension of earth history. The same instruments are also contributing to the finding of earth resources, such as petroleum and deposits of metal ores, that are useful to man.

Geology has become an instrumental science. It ceased many years ago to be the exclusive province of a man roaming the hills with a hammer and a magnifier, although in my prejudiced opinion that is still where the greatest enjoyment is to be found. But great progress in geology is being made with computers, with deep-diving submarines, drilling ships at sea, and with expensive electronic and optical equipment in laboratories. The specialists who are called paleontologists, geochemists, geophysicists, crystallographers, and the like are all varieties of geologists; they are all contributing to an understanding of earth history; they are all "students of the earth"—that is what the word geologist means.

May we earth lovers join together to cherish and protect the earth—protect it from men who seem to want to damage or destroy that which they ought for their own good and for that of their progeny to use with care and wisdom.

APPENDIX

The Properties of Some Common Minerals of the Rocks of the Sierra Nevada

Mineral	*Chemical composition*	*Color*	*Cleavage*
Quartz	SiO_2 (Silica)	Colorless, gray, white.	None
Orthoclase (Potassium feldspar)	$KAlSi_3O_8$ (Potassium, aluminum silicate)	Colorless, pink, white.	2 at 90°: 1 perfect, 1 less perfect
Plagioclase (Sodium-calcium feldspar)	Between $NaAlSi_3O_8$ and $CaAl_2Si_2O_8$ (Sodium, calcium, aluminum silicate)	White, gray, colorless.	2 at nearly 90°: 1 perfect, 1 less perfect.
Muscovite (White mica)	$KAlSi_3O_{10}(OH,F)_2$ (Potassium, aluminum silicate)	Gray, white, pale yellow or green, colorless, transparent	1 highly perfect (micaceous); splits into thin, flexible, elastic sheets.
Biotite (Black mica)	$K(Mg,Fe)_3AlSi_3O_{10}(OH,F)_2$ (Potassium, magnesium, iron, aluminum silicate)	Light to dark brown; typically black.	1 highly perfect (micaceous); splits into thin flexible elastic sheets.
Hornblende (Amphibole group)	$(Na,Ca)_2(Mg,Fe,Al)_5(Al,Si)_3O_{22}(OH)_2$ (Sodium, calcium, iron, magnesium, aluminum silicate)	Light to dark green, black, dark brown.	2 very good at 56° and 124° (roof-top angle).
Augite (Pyroxene group)	$(Ca,Mg,Fe,Al)_2(Si,Fe,Al)_2O_6$ (Calcium, magnesium, aluminum, iron, aluminum silicate)	Dark green to black.	2 at 90°, but seldom visible.
Hypersthene (Pyroxene group)	$(Mg,Fe)SiO_3$ (Magnesium iron silicate)	Light to dark green, brown, black.	2 at 90°, but seldom visible.
Olivine	$(Mg,Fe)_2SiO_4$ (Magnesium, iron silicate)	Light to dark green; yellow to golden by reason of alteration, often with a reddish rim.	2 weak; usually not visible.
Chlorite	$(Mg,Fe,Al)_3(Al,Si)_3O_{10}(OH)_6$ (Magnesium, iron, aluminum silicate)	Light to dark green, mostly very dark.	1 perfect (micaceous); flakes flexible but not elastic.

Luster	*Hardness and [Fracture]*	*Typical Shapes of Crystals or Grains*	*Occurrence*
Glassy (vitreous)	Hard [rough]	Square, hexagonal, elongate rectangular.	In a variety of igneous, metamorphic, and sedimentary rock. Crystals in volcanic rocks and veins.
Vitreous	Hard [rough]	Square to short rectangular.	Crystals in volcanic rocks. Partial to complete crystal form in granitic rocks. Irregular granular in metamorphic rocks.
Vitreous	Hard [rough]	Square, short to long rectangular.	Crystals in volcanic rocks. Partial to complete crystal form in plutonic rocks. Irregular granular in metamorphic rocks.
Smooth, pearly, vitreous	Soft [not applicable]	Hexagonal plate or short rectangular.	Not in volcanic rock, but in metamorphosed silicic volcanic rocks and in silicic plutonic rocks. Abundant in slate and micaceous schist as scaly flakes.
Smooth, shiny, vitreous	Soft [not applicable]	Hexagonal plate or short rectangular.	In a wide variety of igneous and metamorphic rocks.
Vitreous	Hard [rough]	Diamond, elongate rectangular to needle-like.	Volcanic, plutonic, and metamorphic rocks.
Vitreous	Hard [rough]	Square, octagonal, short rectangular.	Andesite, basalt, gabbro, pyroxenite, peridotite. Often preserved in metamorphosed mafic igneous rocks.
Vitreous	Hard [rough]	Square, elongate rectangular.	Andesite, basalt, gabbro, pyroxenite, peridotite. Rare in metamorphic rocks.
Vitreous	Hard [rough]	Diamond, square, short rectangular.	Crystals in basalt, granular aggregates in plutonic igneous rocks. Absent from metamorphic rocks.
Pearly to waxy	Very soft [not applicable]	Hexagonal flakes, aggregates of irregular plates or flakes.	Alteration product of biotite. Fills amygdules in metamorphosed basalt and andesite. In veins, and in slate with muscovite.

Mineral	*Chemical composition*	*Color*	*Cleavage*
Epidote	$Ca_2(Al,Fe)_3Si_3O_{11}$ (Calcium, iron, aluminum silicate)	Pale to dark green, yellow green; less commonly black.	2: 1 perfect, 1 good.
Calcite	$CaCO_3$ (Calcium carbonate)	Colorless, white, gray.	3 perfect.
Dolomite	$CaMg(CO_3)_2$ (Calcium magnesium carbonate)	Colorless, white, gray.	3 perfect; surfaces may be curved.

Luster	*Hardness and [Fracture]*	*Typical Shapes of Crystals or Grains*	*Occurrence*
Vitreous	Hard [rough]	Rhomboidal, elongate rectangular to needle-like.	An alteration product of other minerals such as plagioclase, hornblende, augite. In grains of those minerals and in veins and vesicles. In igneous and metamorphic rocks.
Vitreous, pearly	Soft [not applicable]	Rhomboidal, irregular.	Principal mineral of all limestone and marble, also in veins, amygdules.
Vitreous	Soft [not applicable]	Rhomboidal, irregular.	With calcite in the rock called dolomite or dolostone.

Bibliography

Allen, V. T.
1929 The Ione Formation of California. *Univ. Calif. Publ. Geol. Sci.* 18: 347–448.
Anderson, C. A.
1933 The Tuscan Formation of northern California. *Univ. Calif. Publ. Geol. Sci.* 23:215–76.
Anderson, C. A., and Russell, R. D.
1933 Tertiary formations of the northern Sacramento Valley, California. *Calif. Jour. Mines and Geol.* 35:219–53.
Anderson, T. B., Woodard, G. D., Strathouse, S. M., and Twichell, M. K.
1974 Geology of a late Devonian fossil locality in the Sierra Buttes Formation, Dugan Pond, Sierra City quadrangle, California. *Geol. Soc. Amer. Abstracts with Programs* 6:319.
Axelrod, D. I.
1956 Mio-Pliocene floras from west-central Nevada. *Univ. Calif. Publ. Geol. Sci.* 33:1–322.
1957 Late Tertiary floras and the Sierra Nevada uplift. *Geol. Soc. Amer. Bull.* 68:19–45.
1962 Post-Pliocene uplift of the Sierra Nevada. *Geol. Soc. Amer. Bull.* 73:183–98.
1976 History of the coniferous forests, California and Nevada. *Univ. Calif. Publ. Botany* 70:1–62.
1980 Contributions to the Neogene paleobotany of central California. *Calif. Univ. Publ. Geol. Sci.* 121:1–212.
Bailey, E. H., ed.
1966 *Geology of northern California.* Calif. Div. Mines and Geol. Bull. 190. 507 pp.
Berry, D. T.
1979 Geology of the Portola and Reconnaissance Peak quadrangles, Plumas County, California. M.S. thesis, Univ. Calif., Davis. 87 pp.
Birkeland, P. W.
1963 Pleistocene volcanism and deformation of the Truckee area, north of Lake Tahoe, California. *Geol. Soc. Amer. Bull.* 74:1453–64.

1964 Pleistocene glaciation of the northern Sierra Nevada, north of Lake Tahoe, California. *Jour. Geol.* 72:810–24.

Bond, G. C., and Devay, J. C.

1980 Pre-Upper Devonian quartzose sandstones in the Shoo Fly Formation, northern California—petrology, provenance and implications for regional tectonics. *Jour. Geol.* 88:285–308.

Bonham, H. F.

1969 *Geology and mineral deposits of Washoe and Storey counties, Nevada.* Nevada Bur. Mines Bull. 70. 140 pp.

Brizzolara, D. W.

1979 Geology of the northern halves of the Crocker Mountain and Dixie Mountain quadrangles, Plumas County, California. M.S. thesis, Univ. Calif., Davis. 110 pp.

Burnett, J. L., and Jennings, C. W.

1962 *Geologic map of California*, Olaf P. Jenkins edition, *Chico sheet.* Calif. Div. Mines and Geol. Scale 1:250,000.

Christensen, M. N.

1966 Late Cenozoic movements in the Sierra Nevada of California. *Geol. Soc. Amer. Bull.* 77:163–82.

Clark, B. L., and Anderson, C. A.

1933 Wheatland Formation and its relation to early Tertiary andesites in the Sierra Nevada. *Geol. Soc. Amer. Bull.* 49:931–56.

Clark, L. D.

1960 Foothills fault system, western Sierra Nevada, California. *Geol. Soc. Amer. Bull.* 71:483–96.

1964 *Stratigraphy and structure of part of the western Sierra Nevada metamorphic belt, California.* U.S. Geol. Surv. Prof. Paper 410. 70 pp.

1976 *Stratigraphy of the north half of the western Sierra Nevada metamorphic belt.* U.S. Geol. Surv. Prof. Paper 923. 62 pp.

Clark, L. D., McMath, V. E., and Silberling, N. J.

1962 *Angular unconformity between Mesozoic and Paleozoic rocks in the northern Sierra Nevada, California.* U.S. Geol. Surv. Prof. Paper 450-B, Art. 6:B15–B19.

Compton, R. R.

1955 Trondhjemite batholith near Bidwell Bar, California. *Geol. Soc. Amer. Bull.* 66:9–44.

Condit, C.

1944 The Remington Hill flora. In *Pliocene floras of California and Oregon,* edited by R. W. Chaney, 21–55. Carnegie Inst. Wash. Publ. 553.

Creely, R. S.

1965 *Geology of the Oroville quadrangle, California.* Calif. Div. Mines and Geol. Bull. 184. 86 pp.

Crickmay, C. H.

1933 Mount Jura investigation. *Geol. Soc. Amer. Bull.* 44:895–926.

Curtis, G. H.

1954 Mode of origin of the pyroclastic debris in the Mehrten Formation of the Sierra Nevada. *Univ. Calif. Publ. Geol. Sci.* 29:453–502.

Curtis, G. H., Evernden, J. F., and Lipson, J. I.
1958 *Age determination of some granitic rocks in California by the potassium-argon method.* Calif. Div. Mines and Geol. Spec. Rep. 54. 16 pp.

D'Allura, J. A.
1977 Stratigraphy, structure, petrology and regional correlations of metamorphosed Paleozoic rocks in portions of Plumas, Sierra and Nevada counties, California. Ph.D. diss., Univ. Calif., Davis. 338 pp.

D'Allura, J. A., Moores, E. M., and Robinson, L.
1977 Paleozoic rocks of the northern Sierra Nevada: their structural and paleogeographic implications. In *Paleozoic paleogeography of the western United States,* edited by J. H. Stewart, C. H. Stevens, and A. E. Fritche, 1:395–408. Society of Economic Paleontologists and Mineralogists, Pacific Section, Pacific Coast Paleogeography Symposium.

Dalrymple, G. B.
1964 Cenozoic chronology of the Sierra Nevada, California. *Univ. Calif. Publ. Geol. Sci.* 47. 41 pp.

Devay, J. C.
1981 The petrology and provenance of the Shoo Fly Formation quartzose sandstones, northern Sierra Nevada, California. M.S. thesis, Univ. Calif., Davis. 114 pp.

Devay, J. C., and Stanley, E.
1979 Radiolaria from the Devonian Elwell Formation, northern Sierra Nevada, California. *Geol. Soc. Amer. Abstracts with Programs* 11:412.

Diller, J. S.
1908 *Geology of the Taylorsville region, California.* U.S. Geol. Surv. Bull. 353. 128 pp.

Durrell, C.
1944 Andesite breccia dikes near Blairsden, California. *Geol. Soc. Amer. Bull.* 55:255–72.
1959a Tertiary stratigraphy of the Blairsden quadrangle, Plumas County, California. *Univ. Calif. Publ. Geol. Sci.* 34:161–92.
1959b The Lovejoy Formation of northern California. *Univ. Calif. Publ. Geol. Sci.* 34:193–220.

Durrell, C., and D'Allura, J. A.
1977 Upper Paleozoic section in eastern Plumas and Sierra counties, northern Sierra Nevada, California. *Geol. Soc. Amer. Bull.* 88:844–52.

Durrell, C., and Proctor, P. D.
1948 *Iron ore deposits near Lake Hawley and Spencer Lakes, Sierra County, California.* Calif. Div. Mines and Geol. Bull. 129, Ch. L:67–192.

Ehrenberg, S. M.
1975 Feather River ultramafic body, northern Sierra Nevada, California. *Geol. Soc. Amer. Bull.* 86:1235–43.

Evernden, J. F., and Kistler, R. W.
1970 *Chronology of emplacement of Mesozoic batholithic complexes in California and western Nevada.* U.S. Geol. Surv. Prof. Paper 623. 42 pp.

Ferguson, H. G., and Gannett, R. W.
1932 *Gold quartz veins of the Alleghany District, California.* U.S. Geol. Surv. Prof. Paper 172. 139 pp.

Gianella, V. P.
1936 *Geology of the Silver City District and the southern portion of the Comstock Lode, Nevada.* Univ. Nevada Bull. 30. 105 pp.
1950 Earthquake and faulting, Fort Sage Mountains, California, December 1950. *Seismological Soc. Amer. Bull.* 47:173–77.

Gilbert, G. K.
1917 *Hydraulic mining debris in the Sierra Nevada.* U.S. Geol. Surv. Prof. Paper 105. 154 pp.

Hannah, J. L.
1980 Stratigraphy, petrology, paleomagnetism and tectonics of Paleozoic arc complexes, northern Sierra Nevada, California. Ph.D. diss., Univ. Calif., Davis. 323 pp.

Hietanen, A.
1951 Metamorphic and igneous rocks of the Merrimac area; Plumas National Forest, California. *Geol. Soc. Amer. Bull.* 62:565–607.
1973 *Geology of the Pulga and Bucks Lake quadrangles, Butte and Plumas counties, California.* U.S. Geol. Surv. Prof. Paper 731. 66 pp.
1976 *Metamorphism and plutonism around the Middle and South forks of the Feather River, California.* U.S. Geol. Surv. Prof. Paper 920. 30 pp.
1977 *Paleozoic-Mesozoic boundary in the Berry Creek quadrangle, northwestern Sierra Nevada, California.* U.S. Geol. Surv. Prof. Paper 1027. 22 pp.
1979 Western metamorphic belt north of the North Yuba River, California. *Geol. Soc. Amer. Abstracts with Programs* 11:84.
1981a *Geology west of the Melones Fault between the Feather and North Yuba rivers, California.* U.S. Geol. Surv. Prof. Paper 1226a. 35 pp.
1981b *Feather River area as a part of the Sierra Nevada suture system in California.* U.S. Geol. Surv. Prof. Paper 1226b. 13 pp.

Hill, M.
1975 *Geology of the Sierra Nevada.* Calif. Natl. History Guide 37. Univ. Calif. Press. 232 pp.

Hudson, F. S.
1951 Mount Lincoln–Castle Peak area, Sierra Nevada, California. *Geol. Soc. Amer. Bull.* 652:931–52.
1960 Post-Pliocene uplift of the Sierra Nevada, California. *Geol. Soc. Amer. Bull.* 71:1547–73.

Hyne, N. J., Chelminski, P., Court, J. E., Gorsline, D. S., and Goldman, C. R.
1972 Quaternary history of Lake Tahoe, California-Nevada, *Geol. Soc. Amer. Bull.* 83:1435–48.

Jenkins, D.
1980 Petrology and structure of the Slate Creek ultramafic body, Yuba County, California. M.S. thesis, Univ. Calif., Davis. 164 pp.

Jenkins, O. P., ed.
1948 *Geologic guidebook along Highway 49—Sierran gold belt. The Mother Lode Country.* Calif. Div. Mines and Geol. Bull. 141. 164 pp.

1978 *The great watershed of California*. Angel Press, Monterey, Calif. 41 pp.

Johnston, W. D., Jr.
1940 *The gold Quartz veins of Grass Valley, California*. U.S. Geol. Surv. Prof. Paper 194. 101 pp.

Lindgren, W.
1897 *Description of the Truckee quadrangle, California*. U.S. Geol. Surv. Geol. Atlas, Folio 39.
1900 *Description of the Colfax quadrangle, California*. U.S. Geol. Surv. Geol. Atlas, Folio 66.
1911 *The Tertiary Gravels of the Sierra Nevada of California*. U.S. Geol. Surv. Prof. Paper 73. 226 pp.

Lovering, J. K., and Durrell, C.
1959 Zoned gabbro pegmatites of Eureka Peak, Plumas County, California. *Jour. Geol.* 67:253–68.

Lydon, P. A.
1968 Geology and lahars of the Tuscan Formation, northern California. In *Studies in volcanology,* edited by R. R. Coats, R. L. Hay, and C. A. Anderson, 441–73. Geol. Soc. Amer. Mem. 116.

Lydon, P. A., Gay, T. E., and Jennings, C. W.
1960 *Geologic map of California,* Olaf P. Jenkins edition. *Westwood sheet*. Calif. Div. Mines and Geol. Scale 1:250,000.

MacGinitie, H. D.
1941 *A middle Eocene flora from the central Sierra Nevada*. Carnegie Inst. Wash. Publ. 534. 178 pp.

McJannet, G. S.
1957 Geology of the Pyramid Lake–Red Rock Canyon area, Washoe County, Nevada. M.A. thesis, Univ. Calif., Los Angeles. 125 pp.

McJunkin, R. D., Davis, T. E., and Criscione, J. J.
1979 An isotopic age for Smartville ophiolite and the obduction of metavolcanic rocks in the northwestern Sierran Foothills, California. *Geol. Soc. Amer. Abstracts with Programs* 11:91.

McMath, V. E.
1958 Geology of the Taylorsville area, Plumas County, California. Ph.D. diss., Univ. of Calif., Los Angeles, 199 pp.
1966 Geology of the Taylorsville area, northern Sierra Nevada. In *Geology of Northern California,* edited by E. H. Bailey, 173–83. Calif. Div. Mines and Geol. Bull. 190.

Mathieson, S. A.
1981 Pre- and post-Sangamon glacial history of a portion of Sierra and Plumas counties, California. M.S. thesis, Calif. State Univ., Hayward. 258 pp.

Matsutsuyu, B. A.
1979 Geology of the southern halves of the Crocker Mountain and Dixie Mountain quadrangles, Plumas County, California. M.S. thesis, Univ. Calif., Davis. 169 pp.

Moores, E. M., and Day, H. W.
1984 Overthrust Model for the Sierra Nevada. *Geology* 12:416–19.

Moores, E. M., Day, H. W., and Xenophontos, C.
1979 The Nevadan orogeny, northern Sierra Nevada: an abrupt arc-arc collision. *Geol. Soc. Amer. Abstracts with Programs* 11:118.

Morrison, R. B.
1964 *Lake Lahontan—geology of the southern Carson Desert, Nevada.* U.S. Geol. Surv. Prof. Paper 401. 156 pp.

Pakiser, L. C., and Brune, J. N.
1980 Seismic models of the root of the Sierra Nevada. *Science* 210: 1088–94.

Potbury, S. S.
1937 *The La Porte flora of Plumas County, California.* Carnegie Inst. Wash. Publ. 465:29–82.

Robinson, L.
1975 Geology of the Arlington Formation, Butt Lake area, Plumas County, California. M.S. thesis, Univ. Calif., Davis. 76 pp.

Saleeby, J. B., and Moores, E. M.
1979 Zircon ages on northern Sierra Nevada ophiolite remnants and some possible regional correlations. *Geol. Soc. Amer. Abstracts with Programs* 11:125.

Schweickert, R. A., and Cowan, D. S.
1975 Early Mesozoic tectonic evolution of the western Sierra Nevada, California. *Geol. Soc. Amer. Bull.* 86:1329–36.

Schweickert, R. A., and Snyder, W. S.
1981 Paleozoic plate tectonics of the Sierra Nevada and adjacent regions. In *The geotectonic development of California*, edited by W. G. Ernst, Rubey vol. 1, 182–202. Englewood Cliffs, N.J.: Prentice-Hall.

Speed, R. C., and Moores, E. M.
1980 *Geologic cross section of the Sierra Nevada and the Great Basin along 40° N. Lat., northeastern California and northern Nevada.* Geol. Soc. Amer. Map and Chart Ser. MC-28L. 12 pp.

Strand, R. L.
1972 Geology of a portion of the Blue Nose quadrangle, Plumas and Sierra counties, California. M.S. thesis, Univ. Calif., Davis. 86 pp.

Tuminas, A. C.
1983 Geology of the Grass Valley–Colfax region, Sierra Nevada, California. Ph.D. diss., Univ. Calif., Davis. 302 pp.

Turner, H. W.
1891 Mohawk Lake Beds, Plumas County, California. *Phil. Soc. Wash. Bull.* 11:385–409.
1895 The rocks of the Sierra Nevada. *U.S. Geol. Surv. Ann. Rep.* 14: 435–95.
1896 Further contributions to the geology of the Sierra Nevada. *U.S. Geol. Surv. Ann. Rep.* 17:521–1076.
1897 *Description of the Downieville quadrangle, California.* U.S. Geol. Surv. Geol. Atlas, Folio 43.

Vaitl, J.
1980 Geology of the Cherokee area, northern Sierra Nevada, California. M.S. thesis, Univ. Calif., Davis. 93 pp.

Van Couvering, J. A.
1962 Geology of the Chilcoot quadrangle, Plumas and Lassen counties, California. M.A. thesis, Univ. Calif., Los Angeles. 124 pp.

Varga, R. J.
1980 Structural and tectonic evolution of the early Paleozoic Shoo Fly Formation, northern Sierra Nevada, California. Ph.D. diss., Univ. Calif., Davis. 248 pp.

Wagner, D. L., Jennings, C. W., Bedrossian, T. L., and Bortugno, E. J.
1981 *Geologic map of California, Sacramento quadrangle.* Calif. Div. Mines and Geol. Scale 1:250,000.

Whitney, J. D.
1880 *The auriferous gravels of the Sierra Nevada of California.* Harvard Coll. Mus. Comp. Zool. Mem. 6. 569 pp.

Wilhelms, D. E.
1958 Geology of the eastern portion of the Spring Garden quadrangle, Plumas County, California. M.A. thesis, Univ. Calif., Los Angeles. 71 pp.

Zigan, S. M.
1981 Structure and stratigraphy of the La Porte ophiolite sequence of Sierra Nevada, California. M.S. thesis, Univ. Calif., Davis. 100 pp.

Index

Italicized page numbers indicate where terms are defined; n = note; t = table.

CPSIA information can be obtained
at www.ICGtesting.com
Printed in the USA
BVHW03s1128290318
511895BV00001B/3/P

9 780520 056916